David Zarembka (1943-2020)

From the Editor

I first met David Zarembka at Friends General Conference in 2007 at the introduction of Joy M. Zarembka's book, *The Pigment of Your Imagination: Mixed Race in a Global Society*. I contacted him to volunteer in Kenya leading up to the World Conference of Friends in 2012. I lived for four delightful months in their (Gladys') house in Lumakanda.

David Zarembka published 646 Reports from Kenya from January, 2009, to March 12, 2021. The last 396 of these, beginning with #250 "Perilious times for Kenya's Somalis" October 5, 2013, are posted at the Google Group, "Reports from Kenya." There is a wealth of material in those 646 Reports from Kenya, so we expect to publish more books.

I had been proofreading David Zarembka's Reports from Kenya for several years. Each Wednesday, I woke up to one of his Reports for me to proofread for him to publish on Friday. On Wednesday, March 17, 2021, his email said there would be no Report that week because he was sick, that he had fever and cough. I replied, "Oh dear! That sounds like COVID." The next Wednesday I had a Facebook request from Getry for prayers for David and his wife, Gladys, both of whom were in the hospital. By Thursday when I texted Getry, Gladys had died. The next Thursday, April 1, 2021, David Zarembka died, a great loss for us all!

In 2018 he asked me to help him publish a collection of his reports on small-scale farming in Kenya as a book. I talked him into joining a "Circle of Discernment" that was starting under the Quaker Institute for the Future (QIF) on the subject of Regenerative Agriculture. Four of us were meeting on Zoom to work on a *QIF Focus Book* on agriculture, but that work slowed down and finally stopped in the pandemic. I wish I had published his book in 2018. Now it must be published as a memorial to his devotion to Africa and his extraordinary peace work.

—Judy Lumb
Barranco, Belize
21 December 2021

From His Children

With my father's untimely and unfortunate death, my brother and I have been combing through his life and reflecting on all of the incredible lessons he taught us. We have been reliving our fun childhood in Pittsburgh, archiving photos, and organizing his correspondences. We recently came across a 1964 letter he wrote to his mother while he was in the Peace Corps where he wrote, "I would love to be a farmer."

After our parents met and married in Kenya, they moved to Pittsburgh, PA where they both pursued their education. They were shocked by how difficult it was to procure fresh fruits and vegetables, unlike the ease they had experienced in Kenya. They started going to the "Strip District," area in Pittsburgh near the Allegheny River with a vibrant network of produce wholesalers. They would buy various fruits and vegetables and sell them to their neighbors out of their basement. This became so popular that they eventually opened the "East End Food Co-op," a precursor to Whole Foods, which still operates today. Over the years, Mom and Dad created more food co-ops and wholesaler co-ops. Each week, Dad would travel to Amish farms to buy cheese to sell to others; Tommy and I would trade off weeks in which we would get to accompany Dad to the farm.

Growing up, Tommy and I played in a large backyard filled with a swing set, apple trees, and a grapevine. Over the years, the play area would become less and less as my Kenyan mother slowly turned the grass into a fertile garden and then, into a mini-farm. Dad observed, advised, and ate well. We invited friends over to pick grapes off the vine and our pre-dinner chore included grabbing fresh ears of corn off the stalk or picking beans. By the time the Zarembkas left that house, there was only a small strip of green grass down the middle of the yard, with amazing vegetables, plants, and trees covering the remainder.

Dad got his true wish of being a farmer when he and Mama Gladys moved back to Kenya in 2007. He continued to observe, advise and eat well. They raised animals and grew their own food. They tried and failed and tried and succeeded, as all farmers do. Dad's favorite poem when he was a kid was "Keep A-Goin'" by Frank Stanton and it includes the line, "When the weather kills your crop, keep a-goin!/ Though 'tis work to reach the top, keep a-goin." Dad always encouraged us to persevere and never give up, much as he did throughout his life, pursuing projects and research, big and small. Over the years, Dad's passionate observations allowed him to explore and understand small-scale farming in Kenya. The results are this book. May the wisdom keep a-goin'.

—Joy Zarembka

My dad had boundless curiosity. I remember one time when we were stuck in traffic in the middle of Ohio, my dad decided to follow the exiting traffic--we ended up at the county fair. Or the time, a friend gave me a pair of cross-country skis and with the first snowfall, my dad was cross-country skiing for the first time. Or the time we dropped Joy off for her freshwoman year at Haverford and we continued up interstate 95 all the way to the Bay of Fundy in New Brunswick, Canada. He was utterly amazed at how fast the tide went out. He spent almost an hour standing where the tide was, taking a step to the new new tide line and yelling to me (I got bored in the first five minutes): "Look where it is now!"

My dad was a huge fan of George Fox. When I asked him why, his response was that he liked the way when George Fox wanted to learn something about the Native Americans, he would just go. My dad was the same way. If he wanted to learn something--he would go. It is not surprising that he would live his final years in a small town almost seven hours away from Nairobi.

For him, the world was a place to be discovered. And he loved to share his discoveries. His blog, "Reports from Kenya", is all about what he saw, experienced and, oftentimes, formed an opinion on. Agriculture, and small scale farming in particular, always intrigued him. The material for this book is from his posts.

Adjusting to a reality where my dad is not an email or call away is difficult. I am thankful to Judy for spearheading this project and allowing the fruits of my father's boundless curiosity to continue to grow."   —Tommy Zarembka

Small-Scale Farming in Kenya

David Zarembka

Producciones de la Hamaca 2022

Published by *Producciones de la Hamaca*, Caye Caulker, Belize
<producciones-hamaca.com>

ISBN: 978-976-8273-30-7 (paperback)
ISBN: 978-976-8273-31-4 (e-book)

Small-Scale Farming in Kenya is composed of "Reports from Kenya" by David Zarembka <posted at the Google Group of the same name>.

This book was printed on-demand by Lightning Source, Inc (LSI). The on-demand printing system is environmentally friendly because books are printed as needed, instead of in large numbers that might end up in someone's basement or a dump site. In addition, LSI is committed to using materials obtained by sustainable forestry practices. LSI is certified by Sustainable Forestry Initiative (SFI® Certificate Number: PwC-SFICOC-345 SFI-00980). The Sustainable Forestry Initiative is an independent, internationally recognized non-profit organization responsible for the SFI certification standard, the world's largest single forest certification standard. The SFI program is based on the premise that responsible environmental behavior and sound business decisions can co-exist to the benefit of communities, customers and the environment, today and for future generations <sfiprogram.org>.

Producciones de la Hamaca is dedicated to:

—Celebration and documentation of Earth
    and all her inhabitants,
—Restoration and conservation of Earth's
    natural resources,
—Creative expression of the sacredness of
    Earth and Spirit.

Contents

Preface

Small-scale farmers make up the large majority of the farmers of the world. They produce an estimated 70 to 80 percent of the food in the world. They will be the ones who will increase food production by at least 50 percent to feed the assumed 10 billion people living on earth in 2050.

By small-scale farmers I mean farmers who cultivate one-tenth of an acre up to around twenty acres of arable land. I do not like the term "subsistence farmer" because usually even the subsistence farmer produces more than he/she can eat and sells/gives away surpluses. Likewise the term "peasant" has come to have negative connotations (definition: "a poor smallholder or agricultural laborer of low social status") so I don't like that one either.

Let me clearly state my thesis: It is a myth that large commercial, agriculture based on large farm equipment, fertilizers, insecticides, pesticides, mostly under corporate management will produce enough to feed the world. This is their claim, but it is a false one. Since the agricultural revolution began a few thousand years ago, the world has always been fed by the efforts of small-scale farmers. Moreover there has always been and is even today more than enough food to adequately feed all the people in the world. When there has been starvation and famine, as recently in South Sudan and now in Yemen, it is always due to war and/or mis-governance. The problem is that poor people usually then don't have the money to buy food that is available, frequently at an exorbitant price as the middlemen cash in.

That this is contrary to the conventional wisdom of the educated world is the purpose for writing this book.

My Qualifications

What are my qualifications for writing these posts? I have no academic qualifications whatsoever in agriculture. This is a resounding positive factor. I have not been indoctrinated in what farmers — meaning commercial agribusiness farmers — ought to be doing. I have not been indoctrinated by the large scale commercial farming philosophy promoted by most agricultural schools. I have seen so many articles and books that discuss the small-scale farmers with the attitude that, since they are not listening to and following the advice of the "experts", it is because they are too conservative, uneducated, or at worse "backwards."

My approach has been quite different. I look and see what actual small-scale farmers do. I ask questions and find out what they do and often why they do what they do. If the farmers do not follow what the experts advise, I consider the "experts" to be the ones who are wrong. No innovation, no improvement can occur unless the clients, the recipient farmers, are willing to adopt it. If they do not accept the changes, there must be some reasons and those reasons must be addressed. If they can't be addressed, then the innovation is moot.

I have lived mostly in four different places in Tanzania and Kenya (*see map on the next page*)

1) In 1965 and 1966 I lived for 15 months in Muzenzi refugee camp in Ngara District of Tanzania near the Burundian and Rwandan borders. At this time the local population was still using shifting "slash and burn" agriculture where they would clear and burn a plot and cultivate it for a few years until the fertility declined. When this happened, they would shift to and clear another plot of the forest and bush. Population density used to be very low and land was always available to clear. This is no longer the case, so farmers have had to develop other methods of maintaining their soil fertility.

2) From 1966 to 1967 I was assigned by the US Peace Corps to a government sponsored project called Rwamkoma Settlement Scheme near the town of Musoma on the east side of Lake Victoria. In another section called "What not to Do" I'll relate my experiences there.

3) From 1968 to 1970, I lived for two years and four months in the Mua Hills, Machakos District, east of Nairobi. Here I was the founding principal of what is now the Mua Hills Girls Secondary School. The school was one of the first few high schools in Kenya to offer studies in agriculture.

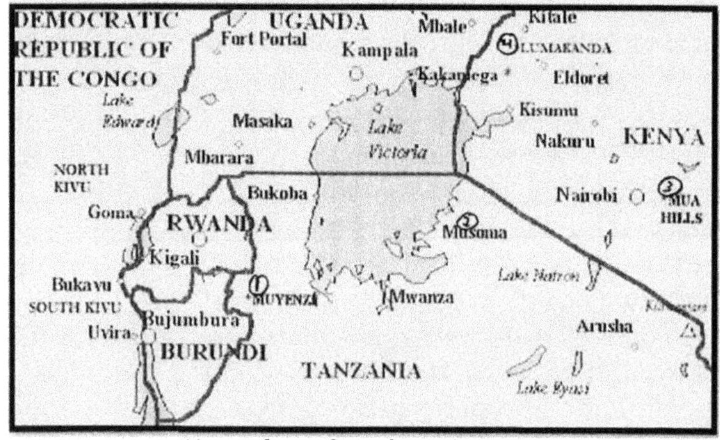

Map of my four locations

4) I married Rodah Wayua at that time and we bought a four-acre farm which was later increased to a total of six acres, ten kilometers (six miles) from the school.

5) From 2007 to 2021, my second wife, Gladys, and I have lived in Lumakanda, Kakamega county, in western Kenya. Currently we have a three-quarter acre homestead plot and we also cultivate two acres nearby that belong to Gladys' son, Douglas. We are small-scale farmers. Gladys is the person responsible for the agricultural work because I frankly don't know all the myriad details that are needed to farm success-fully. On the other hand, I have a first-hand participant's view of how various agricultural activities are done.

All four of these sites have commonalities. First all are between 1500 and 1825 meters (5,000 and 6,000 feet) above sea level. This means that there is adequate average rainfall from 75 to 125 centimeters (30 inches to 50 inches) per year. All the sites are on fairly fertile volcanic soil. As a result of these attri-butes all these sites are relatively highly populated, although Ngara District in 1960s was not nearly as thickly populated as it is now.

I have also made numerous trips to Uganda, Rwanda, and Burundi. I have kept my eyes open and asked questions.

In total I have therefore had eighteen and a half years of observing small-scale farmers in East Africa, mostly in Kenya.

The Man with a Hoe

My grandfather, Frank Zarembka, was of Polish peasant background. The word "peasant" is usually used pejoratively as it means "a poor smallholder or agricultural laborer of low social status (chiefly in historical use or with reference to subsistence farming in poorer countries)." There is a long history of degrinating such farmers. Unfortunately I think this has a major relevance to the present political climate in the United States. Hilary Clinton may have lost the 2016 election with her description of Trump supporters as "deplorables"; this elitist attitude is one of the reasons that Trump and the Republicans are popular in rural America.

When I was a child, my Mom had a copy of the book, 101 Famous Poems. Among the "famous" poems was this one (stanzas one and three) by Edwin Markham, which was written in 1898.

The Man with the Hoe

Bowed by the weight of centuries he leans

Upon his hoe and gazes on the ground,

The emptiness of ages in his face,

And on his back the burden of the world.

Who made him dead to rapture and despair,

A thing that grieves not and that never hopes.

Stolid and stunned, a brother to the ox?

Who loosened and let down this brutal jaw?

Whose was the hand that slanted back this brow?

Whose breath blew out the light within this brain?

What gulfs between him and the seraphim!

Slave of the wheel of labor, what to him

Are Plato and the swing of Pleiades?

What the long reaches of the peaks of song,

The rift of dawn, the reddening of the rose?

Through this dread shape the suffering ages look;

Time's tragedy is in the aching stoop;

Through this dread shape humanity betrayed,

Plundered, profaned, and disinherited,

Cries protest to the Powers that made the world.

A protest that is also a prophecy.

Since the Nineteenth Century in American and European culture there have been attacks on the small-scale farmer. As a result, there is a cultural zeitgeist against these farmers with the need to glorify large-scale commercial farming. This cultural negativity is part of the approach of educated people who consider themselves superior to the small-scale farmers. This then leads to the disastrous mistake after mistake of promoting large mechanized agricultural schemes.

Let me give some counter examples from the United States. I have a friend named Karl Meyer who lives in Nashville, TN. He bought a run-down house in a declining section of the city and rehabbed it. He now grows all his food in the front and back yard of his house. The Nashville city government took him to court over the gardens in his front yard as it was expected to be green grass only. He persevered. Think of all the front and back yards in the United States that could be used for growing food. Consider all the fertilizer, pesticides, and weed-killers put on that grass and the pollution from gas lawnmowers that have to keep that grass short. On the other hand, I have many friends who went "back to the land" in western Massachusetts, Maine, and Virginia. One can read *Compact Farms: The Proven Plans for Market Farms on 5 Acres or Less* by Josh Volt (the son of long-time Friends Committee on National Legislation's Executive Secretary, Joe Volt).

From 1971 to 1985 and then from 1988 to 2007 when we moved to Africa, I frequently traveled between first Pittsburgh and St. Louis (where I grew up and my parents lived) and then Washington, D.C., and St. Louis. This is the American corn belt, but the corn grown in the US is not fit for human consumption, unlike the "maize" that is grown in Kenya and eaten by people. The US cornbelt is a "wasteland" of corn and soybeans. There are hardly any people. There are small declining towns here and there between the large cities.

About 40 percent of this corn is turned into ethanol to be blended with gasoline and much of the rest is fed to pigs, poultry, and cattle. It takes 26.5 pounds of corn to produce one gallon of ethanol. This could feed a Kenyan for a month. One of the best methods of feeding the world's population would be a drastic reordering of American agriculture. The massive environmental degradation from this commercial, corporate farming is another concern.

Imagine if the whole US corn belt, some of the most fertile, well-watered land in the world, was full of small farms growing food for people. As we drive through the monotonous corn/soy fields, I often say to Gladys, "What would some good Kenyan farmers do with all these fields? Wouldn't they produce so much more food and repopulate the countryside with farms, communities, businesses, and a lively social society?" Gladys agrees. Do you?

Gladys and Dave at their homestead in Lumakanda, Kenya

CHAPTER ONE: The Destruction of Small-Scale Farming

Country people leaving a village due to enclosure in 18th Century England. (Mary Evans Picture Library)

In the last five hundred years the land of small-scale farmers has been confiscated by elites frequently and forcefully. The rationale has been due to the negative characterization of the small-scale farmer as uncivilized, stupid, unproductive, and backward. These have been the justifications for the elites—whether large landowners, government officials, educated professionals, or others—to seize the small-holder farms.

This began in England in the sixteenth century with the Enclosure Movement. Its climax was 1776 to 1825 when Parliament passed over four thousand Acts of Enclosure. The process of enclosure was often accompanied by force, resistance, and bloodshed and led to large numbers of landless people who became the exploited work force for the industrial revolution in England. Historian Raymond Williams summarized the Enclosure Movement as "concentrated ownership of [land] in a small minority of the population. These 'lawful' enclosures also dispossessed millions of citizens, swept away traditional ways of life, and forcibly introduced the new economy of industrialization, occupational specialties and large-scale production." (Raymond Williams, "Enclosures, Commons and Communities," in *The Country and the City* (Oxford University Press, 1973, pp. 96-107.)

This also led to many English people migrating to the United States, Canada, Australia, New Zealand, South Africa, and other places where the English created large estates by expropriating land from the original inhabitants. As is well known, but usually ignored, the English entered the eastern United States and dispossessed the Native American small-scale farmers of their land, pushing them farther and farther west as they continued to take over their lands.

A 1911 ad offering "allotted Indian land" for sale.

A family starves in their own yard. Ukraine, 1933. Wikimedia Commons picture. This is one of the rather less gruesome pictures in this collection.

A starving peasant family, victims of the disastrous economic policies of the 'Great Leap Forward' leading to the 'Great Famine' (1958 - 1961).

Sohail Khan, a farmer from Chitora village in Raisen district, Madhya Pradesh, India. (Photo: Pankaj Tiwari). Is this what thickly populated India needs? A large field with a tractor with no houses, no people, and hardly any trees?

There were other massive expropriations of land. Joseph Stalin ordered the collectivization of agriculture in Russia and Ukraine. The 1932-33 Ukrainian famine due to this collectivization policy led to the deaths of between 3.3 million and 7.5 million Ukrainians. Including the deaths in Russia itself, the total may have reached 12 million.

Mao Zedong's Great Leap Forward in China from 1958 to 1962 was another government-sponsored policy to turn the small-holder Chinese farms into large collectives. This led to a major famine in China, which led to the deaths of between 18 million and 55.6 million people.

These attacks on the land and livelihood of small-scale farmers are not old history. They continue today, perhaps with less violent methods. The green revolution in India is a good example of the current means of forcing people off their land. The green revolution that began in the 1970s quickly increased yields in parts of India. This led to surpluses in food crops with a corresponding drop in the price of food. At the same time, land became more valuable since it could grow the crops. Since only farmers with large land holdings qualified for loans for the expensive seeds, fertilizer, chemicals, and sometimes irrigation needed for the Green Revolution to succeed, the small holder farmers were unable to make ends meet.

When they went into debt and couldn't pay off the extremely high interest rates charged by the moneychangers, their farms were possessed for non-payment. Some estimates in the major growing areas of India calculate that 30 percent of the farmers were moved off their land. They could then live precariously as day laborers when work was available or move to the vastly over-crowded cities. Suicide among these farmers became a national epidemic. The idea was that this surplus labor, like in England during the Enclosure Movement, would produce the people power for industrialization. This did not happen, so the cities and countryside are teeming with poverty-stricken destitute families.

Lord Delamere's extensive Soysambu estate.

British Concepts of Land Ownership to Kenya

While the first enclosures in England and Wales started in the 13th century, over 5,200 individual enclosure acts were passed between 1604 and 1914 by the British parliaments that enclosed 6.8 million acres of land formerly owned in common by the community. This is about 20 percent of the land area of England and Wales. Whoever lived on the land was removed or made tenants on the land they formerly lived on and farmed. Those people, disposed of their land, responded vigorously, which led to many conflicts and deaths as the enclosures were enforced by the armed power of the state. These are the people who ended up in poverty in the towns of England, and who also immigrated to the US, Canada, Australia, and elsewhere. The result is that the vast majority of land in England and Wales is now owned by a small group of landlords. Twenty-five thousand land owners, less than one percent of the population, control more than 50 percent of the land.

In 1944 George Orwell wrote on enclosure:

"Stop to consider how the so-called owners of the land got hold of it. They simply seized it by force, afterwards hiring lawyers to provide them with title-deeds. In the case of the enclosure of the common lands, which was going on from about 1600 to 1850, the land-grabbers did not even have the excuse of being foreign conquerors; they were quite frankly taking the heritage of their own countrymen, upon no sort of pretext except that they had the power to do so." <alexpeak.com/twr/oateotc>.

Thirty percent of the land in England and Wales is owned by the aristocracy and gentry. This may even be an underestimate, as the owners of seventeen percent of England and Wales remain undeclared at the official Land Registry as many of these estates have remained in their families for centuries.

How did all of this translate to Kenya when the British colonialists arrived? At that time there were no title deeds in the country and people lived under the customary land uses.

Primogeniture is the system where the oldest son inherits the total property of the family. The purpose of this is to keep land from being divided up. Primogeniture in England meant that there were many younger sons who had no estates at all. These younger sons were the ones who came to Kenya and saw all the wonderful Kenyan highlands with a moderate climate. They brought ideas from England; they wanted large estates.

As with the Enclosure Movement in Britain, they had no consideration for the Africans who inhabited the land. They pushed them out to the more marginal areas or allowed them to remain as squatters to work on their farm, but with no legal right to the land that they formerly used. There are about 20 million acres of arable land in Kenya and the settlers seized about 10 million of these acres or half of the best land.

For example, Lord Delamere bought two estates at a nominal price from the colonial government, one of

Old picture of Soysambu workers and cattle.

100,000 acres which he later sold to cover his debts and the other, Soysambu, originally 80,000 acres. A one-hundred-thousand-acre estate is 156 square miles or an area with 12.5 miles on each side. You can still see this second Soysambu estate on the road from Nairobi to Nakuru as it stretches for miles along the road.

Some of the best land was around Mt Kenya. Nairobi was established at the southern end of this area. This area was the home of the Kikuyu, the largest tribe in Kenya. As with the Nandi, the Kikuyu were pushed off much of the land into soon overcrowded areas called "reserves." These injustices led to the Mau-Mau rebellion (1952-1960) in Kenya, mostly in the Mt. Kenya region.

Starting during the Mau-Mau Rebellion, the British colonials began consolidating land in the Kikuyu reserves and giving the new "owners" title deeds. Most of the best land went to Kikuyu supporters of the colonial government because many of the Mau-Mau rebels were in prison with many of their wives, children, and elderly in concentration camps. The end result was that many of the Kikuyu ended up with no legal title to any land or a plot that was much too small and marginal to farm. These disinherited Kikuyu migrated to other parts of Kenya, which is the source of much of the tribal conflicts that still persist more than 50 years after independence.

When land is grabbed in Africa, any people living on that land are moved off by the government involved.

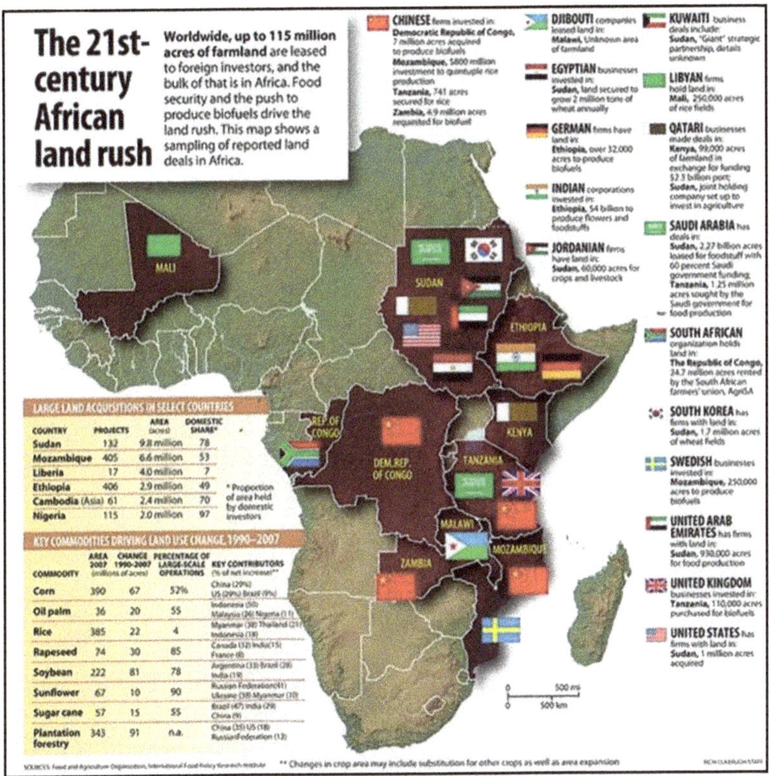

LARGE LAND ACQUISITIONS IN SELECT COUNTRIES			
COUNTRY	PROJECTS	AREA (acres)	DOMESTIC SHARE*
Sudan	132	9.3 million	78
Mozambique	405	6.6 million	53
Liberia	17	4.0 million	7
Ethiopia	406	2.9 million	49
Cambodia (Asia)	61	2.4 million	70
Nigeria	115	2.0 million	97

* Proportion of area held by domestic investors

KEY COMMODITIES DRIVING LAND USE CHANGE, 1990–2007				
COMMODITY	AREA 2007 (millions of acres)	CHANGE 1990-2007	PERCENTAGE OF LARGE-SCALE OPERATIONS	KEY CONTRIBUTORS (% of net increase)**
Corn	390	67	52%	China (29%) US (29%) Brazil (9%)
Oil palm	36	20	55	Indonesia (50) Malaysia (26) Nigeria (11)
Rice	385	22	4	Myanmar (38) Thailand (21) Indonesia (18)
Rapeseed	74	30	85	Canada (32) India (15) France (8)
Soybean	222	81	78	Argentina (33) Brazil (28) India (19)
Sunflower	67	10	90	Russian Federation (61) Ukraine (38) Myanmar (10)
Sugar cane	57	15	55	Brazil (47) India (29) China (9)
Plantation forestry	343	91	n.a.	China (35) US (18) Russian Federation (12)

SOURCES: Food and Agriculture Organization, International Food Policy Research Institute   ** Changes in crop area may include substitution for other crops as well as area expansion

Here is a map and chart of the major land grabbing happening in Africa as of 2007 <newscastmedia.com/blog/wp-content/uploads/2012/07/land-rush.jpg >

Sometimes the people are offered compensation, but even when this happens, it frequently never arrives. If a farmer and his family have been living on a piece of land for generations, what is the value of that land to him since any money he/she might receive is quickly used up? The family joins the long list of landless people eking out an existence with day labor.

Kenya is a large country, but only 20 percent of the land is arable with the rest being arid and semi-arid. There is somewhat more than 20 million acres of arable land. About half of this arable land, 10 million acres, were mostly allocated to British settlers and companies

Large Scale Farming in Kenya

A large international tea estate near Kericho, Kenya. This land was taken from the Kipsigis under a 99-year lease which has just been renewed for another 99 years. The tea companies control 800,000 acres of land and employ 60,000 workers. Tea is Kenya's largest export crop.

in the so-called "White Highlands." Early settlers like Lord Delamare were allocated up to 100,000 acres each. This land allocated to the settlers was usually the best land and that closest to the railroad for easy imports and exports.

The seizure of this land from the Kenyans was done by brute force. When the British were allocated land, the inhabitants on the land were either pushed off or made squatters/workers on the land that formerly had been theirs. There was resistance by the Kenyans to this encroachment.

For example, the Nandi, the tribe that lives just south and east of us in Lumakanda, resisted the longest between

1890 and 1906. The British Army attacked the Nandi. They conducted what would now be considered genocide. Please do not buy into the British myth that they were "benign" colonialists as they were just as brutal as the Germans, French, Belgians, or Italians. Not only did they kill the Nandi warriors, but they also burned their houses and farms, killed their cattle, goats, and sheep, and laid waste to the countryside. Many women and children starved to death after their methods of sustenance were destroyed. The Nandi resistance ended, when on October 19, 1905, the British commander, Col Richard Meinertzhagen, invited the Nandi leader, Koitalel Arap Samoei and his advisors to a meeting to discuss a truce. When they arrived, rather than discuss peace, Meinertzhagen killed Arap Samoei and those with him in cold blood. This brutal British history of expropriation of land by force also occurred in South Africa, Zimbabwe, and to a lesser extent in Zambia and Tanzania.

At Kenyan independence in 1963, elderly Kenyans could remember when their lands were dispossessed; younger people were also told the history by their parents and grandparents. It is therefore no wonder that the Mau Mau rebellion from 1952 to 1960 occurred mostly among the Kikuyu north of Nairobi where much of their very fertile, well-watered land had been seized by the British settlers.

With independence many Kenyans thought that their confiscated lands would be returned to them as the British settlers would be removed from their estates. In order to satisfy this demand, the British and Kenyan governments agreed to buy out British settlers and allocate their farms in small plots of around 20 acres to African small-scale farmers. This was called the "Million Acre Settlement Scheme." Lumakanda where we live is one of those areas where large farms were divided up so we live in an area of small-scale farmers.

This is the British settler's house in downtown Lumakanda. It is now used for stores and offices.

Yet there was still nine million acres of British farms which were not divided up. These were to be sold on a willing seller/willing buyer basis. Most of the British settlers, after their repressive rule in Kenya, panicked and sold out cheaply. The buyers, though, were the new Kenyan elite who had cooperated with the British and had jobs and businesses that gave them the funds necessary to buy out the British farmers. Much of this land therefore just changed from British settlers to Kenyan settlers and the old wounds were not addressed. This is part of the reason for the violence during elections in Kenya as the dispossessed again raise the issue of their lack of land. For example, in the post-election violence after the 2007 election, the main issue near us was that the Nandi felt that the Kikuyu had invaded their land. The violence involved Nandi warriors against new Kikuyu settlers and those whose ancestors had come to the area when they worked on the British settlers' farms.

This is a picture of part of the Del Monte Company's pineapple plantations. The company controls a total of 26,000 acres in two counties north of Nairobi. Their canned pineapples are one of the leading Kenyan exports. Their 99-year lease has ended, but there are still (2021) controversies about the distribution of that land <businessdailyafrica. com/bd/news/counties/murang-a-residents-denied-7-447-acres-del-monte-land-3571712>.

The land issue due to the violent expulsion of the original inhabitants still festers actively in Kenya. Those large farms, held now by wealthy Kenyans and international companies, can easily be seen on the way to Nairobi by those many small farmers who own only a few acres. When elections come, local politicians looking for votes tell their tribesmen to "clean the fields of weeds" (meaning those Kenyans, particularly the Kikuyu, who did not originally live in the area). Violence then ensues.

In my opinion, many of those large farms should be bought up and divided into small-scale farms as was done after independence with the Million Acre Settlement

Scheme. As I have indicated, the result will not be only more equitable, but after settlement by small-scale farmers, the land will be more productive than it is now.

Please do not assume that that this dispossession of small-farmers' land is only history. Yesterday (2018) I received a report from Uganda where more than 1000 people from 35 villages have been evicted. Here is an excerpt from the report, "The community members were given an ultimatum of three hours to vacate the land amidst beatings and shootings from over 300 police and army officials. People's houses were demolished, gardens destroyed and at least 27 people lost their lives, women and children inclusive" <solidarityuganda.org/stop-land-evictions-in-kyangwali>.

What Not to Do, a Personal Experience

In 1966 the US Peace Corps assigned me to Rwamkoma Settlement Scheme in northern Tanzania just east of Lake Victoria. In the enthusiasm of independence, Tanzania decided to jump into modern commercial farming with the establishment of a number of "settlement" schemes. The idea developed by "experts" was that farmers would move together into a compact village with schools, health clinics, piped water, and (maybe) electricity. The schemes would be mechanized so that tractors plowed the fields for the settlers. Each one would be given acreage to grow food crops and a commercial crop to repay the debts from joining the scheme and reimbursements for plowing, seeds, fertilizer, and other costs.

The Rwankoma Settlement Scheme grew cotton as its cash crop. People were given two acres for food crops and two more for cotton. These were marked off in large fields. Each settler was given a small plot in the village where they could build a house and grow

A current picture of the countryside in Musoma District.

vegetables and minor crops. I was posted to the settlement scheme and in time I became the cooperative secretary who recorded the cotton harvests of each settler. Consequently, I had first-hand knowledge of what each settler was producing.

The scheme was a disaster.

First, the number of settler slots were not filled because the local people did not want to live in villages and cultivate fields outside their own plot. Those who were already living there all kept their traditional homesteads and came to the scheme in the hope that there might be some advantages for them.

Second, if the settler grew two acres of good cotton, all the proceeds would be taken to pay for the plowing, seeds, and fertilizer plus the considerable overhead of developing the scheme and paying the settlement scheme staff (manager, assistant manager, coop secretary, mechanic, and tractor drivers).

Third, at that time there were few tractors in Musoma District and, whenever a tractor broke, spare parts were hard to get as they sometimes had to be brought all the way from Dar es Salaam, the capital over 1000 kilometers (625 miles) away over poor roads. As a result the tractors were often out of commission with the result that plowing was often late.

Fourth, due to its altitude of about 1525 meters (5000 feet), cotton growing was marginal and conditions had to be perfect for the farmer to obtain a good yield. Even if he/she did, the yield would just cover the costs so that the settler, after doing the work of planting, weeding, and harvesting the cotton, would remain with no cash income. So what would be the advantage to the farmer from growing the cotton?

Fifth, the experts looked at the average rainfall which was sufficient for cotton, but ignored the variability in the rainfall patterns. One year I was there when rainfall was considerably below normal, the farmers were unable to pay off their debts and had to carry them forward to the next year. As a result all the farmers had considerable debts that they couldn't pay off.

Sixth, there was considerable corruption. Farmers themselves would carry some of the cotton off the scheme and put it with their homestead cotton and take it to the ginnery to be paid cash, therefore getting at least something for their efforts. More important was the fact that the manager of the scheme was stealing in every method he could. He sold seed, fertilizer, diesel, and other inputs and pocketed the proceeds. At night he took cotton from the settlement store and sold it the next day at the ginnery. And so on.

It was clear to me that the scheme was unviable. So I wrote up an analysis of the economics of the scheme and sent it to one of my contacts in Dar es Salaam. I don't know if this had any effect on the scheme as I will describe below, but I did see the paper cited in a study of the Tanzanian settlement schemes.

The situation came to a head for me when the manager went on vacation. He gave the key to the safe to the assistant manager, but clearly kept another one for himself. He counted the money with the assistant manager that was in the safe, but, after he left, when the assistant manager opened the safe, he found all the cash missing. So we went to the local sub-chief to report this theft plus the other methods of embezzlement that the manager was doing. I wrote a letter to the Peace Corps concerning these developments. As soon as they received my letter, they removed me from the settlement scheme. I will have to admit that I was happy to leave since I didn't really want to be part of a sinking enterprise.

By this time in late 1967, relations between the United States and Tanzania had deteriorated. It seems so puerile now. In 1964 Tanganyika joined with Zanzibar to form the new country of Tanzania. Tanganyika had recognized West Germany and Zanzibar had recognized East Germany. President Julius Nyerere decided that Tanzania would then continue to recognize both West and East Germany. This made Tanzania an enemy of the United States. As a result the Peace Corps was being removed, so I was transferred to Kenya and finished out the last eight months of my tour there.

About a month after I left, some of the settlers attacked the manager one night and told him that, if he did not leave the scheme, they would kill him. He left the next day. Within a year, the Tanzanian Government realized that the Rwamkoma Settlement Scheme was a failure and turned it into a beef cattle scheme with few or no settlers. Within the next few years all the other capital intensive settlement schemes were closed.

President Nyerere though, didn't learn. He continued to promote the policy of "villagization" where the farmers would be consolidated into villages, but without the costly overhead that the settlement scheme had. This policy was one of Nyerere's big failures since the local people, who had never been consulted, preferred to live in their own homesteads with their fields adjacent to their houses.

What are the take-aways from this experience?

- The "experts" don't have sufficient knowledge of local conditions to develop viable agricultural development.

- A top down agricultural program will not work.

- The local farmers must be consulted in any innovations and these innovations will only succeed if the farmers see definite benefits.

- Changes must be in accord with local culture. Villagization was contrary to local culture in Tanzania.

- The timeline for agricultural development cannot be compressed into a set scheme. It will take years of experimentation and fine tuning before it will be successful.

- "The best laid schemes of mice and men oft go astray."

- Commercialization of agriculture as practiced in the United States and elsewhere is not viable in a small-scale farming community context.

What Not to Do, Two Examples

There are many cases of grand agricultural schemes in East Africa that promised the moon, but ended in collapse. I will only cover two — Dominion Farm and the Galana Kulalu Irrigation Scheme.

Dominion Farm was started in 2002 by an American capitalist named Calvin Burgess. His fortune was made in real estate. He leases eleven properties he has developed to the US Federal Government. He also is reported to own some private prisons. He leased 17,050 acres in Yala swamp in western Kenya near Lake Victoria to grow rice, tilapia, and later sugar cane.

American mechanization on the Dominion Farm's rice fields. Notice the small-scale farms on the hill in the distance.

This is a quote from the Dominion Group's website: <domgp.com/agriculture.html>. "At full production, the farm will help enable this country to reduce dependence on imported food, will serve as a demonstration of productive farming practices and will return a profit to Dominion. The company is committed to the enrichment of the local population through decent employment, out-grower contracting and the support of schools, clinics and emerging community initiatives." His method was pure American: mechanized, irrigated farming in the middle of a thickly populated area of small-scale farmers.

Fifteen years later in November 2017, the scheme collapsed with Burgess owing 350 million shillings ($3,500,000) in salary arrears plus substantial other debts to Kenyan suppliers. What happened?

Burgess claims that he was unable to continue because of Kenyan corruption and then political opposition. The political opposition developed because the project had lost the support of the local population. None of the promised community development happened. Local people whose houses were possessed for the project were given 6,000 shillings ($60) for their home. Women were paid 200 shillings ($2.00) per day to shoo away the birds eating the rice. Local farmers claimed that the aerial spraying of herbicides (really, is this what Kenya needs?) drifted onto their land and made their animals sick. 2017 was an election year and politicians, including Raila Odinga who comes from this area, withdrew any support that they previously had for the project. I suspect, though, that the major reason was that Burgess was pouring money and more money into the project with no possibility of making a profit.

The lesson is that a wealthy "can-do" capitalist who believes in American exceptionalism ("the theory that the peaceful capitalism of the US constitutes an exception to the general economic laws governing national historical development") is doomed to failure. A healthy dose of humility and skepticism needs to be a major part of any large-scale agricultural scheme.

Only 20 percent of the land area of Kenya is suitable for farming because it is dependent on rain-fed agriculture. Kenya often has a food deficit and has to import food from other countries, including neighboring Uganda and Tanzania. One of the solutions to this problem is irrigation. One of President Uhuru Kenyatta's mega-projects, the "Galana Kulalu Irrigation Scheme," involved irrigation of 1,000,000 acres in an arid area near the coast with water from two nearby rivers to produce, along with other crops, forty 200-lb

Irrigation at the Galana Kulalu Irrigation Scheme

bags of maize per acre. If this happens, it will double the Kenyan annual maize harvest.

In a good year Kenya is nearly self-sufficient in maize. It is unclear where all this extra maize would be sold since the price of maize in Kenya is well above the world price, even that of Uganda and Tanzania. Would its success then put the commercial maize farmers and many small-scale farmers out of business?

This was announced in 2014 and 15 billion shillings ($146,000,000) was allocated to initiate the project's first 10,000 acres. Only 5000 acres were planted that year and the harvest was only 10 bags per acre, well below that from rain-fed agriculture where I live upcountry. The next year 23 bags per acre were harvested from 5145 acres. Then the third year (2017) 31 bags per acre were harvested on only 3000 acres. This was all done by the government itself. Although the concept was that this farming would be a public/private partnership where companies would be licensed to cultivate allocated acreage, after initial interest, it seems there are no private partners willing to invest in this project. This year (2018) there has been little news on the scheme and it is probably dying a natural death.

It is so easy to sit in the armchair and multiply acreages and mythical harvests to make grandiose schemes seem viable. What if all the funds sent for this scheme had been used to support small-scale famers; for example, by subsidizing fertilizer for their maize crop?

My conclusion is that Africa and the world is not going to be fed by grandiose agricultural schemes based on the US model of large, highly mechanized, capital-intensive production. Much better results are going to be had by supporting the vast majority of small-scale farmers to increase their production, shortage, and marketing.

CHAPTER TWO
The Small-Scale Farm

This is a picture of the homestead of the late Rezpa Sabatia, Gladys' cousin, who lived down the road from us. She and her husband, the late James Sabatia, were one of the original settlers in Lumakanda in the early 1960s. There is a corrugated iron building on the right among the trees that was their carpentry shop, James' business to support the family and invest in the farm. They also have the town's slaughter house on their property where one or two cows are slaughtered each day. They have never sold any of their 20 or so acres, but each of the three sons was given about two acres as their homestead where they have built their family house. Nonetheless the rest of the farm is still cultivated as one unit.

One of the misconceptions about small-scale farming is that it should be an economic activity that will allow the farmer (meaning, a man) to grow enough food to feed his family and have sufficient surplus to sell to support a decent lifestyle. This includes funds for housing, clothes, food such as sugar that needs to be bought or when there is a drought, schooling, medical expenses, funerals/weddings, and so on. Therefore, if the farmer or someone in his family has to secure paid income outside the farm to cover some of these expenses, it is considered a failure.

Since twenty acres was what was considered sufficient for a self-sustaining farmer, for the One Million Acre Settlement Scheme in Kenya, about twenty acres were allocated to each settler family. In the beginning the settlers were not allowed to sell or divide any of their acreage, but, since this was so contrary to custom and need, it soon fell by the wayside. Some of these farmers have sold off some or even much of their acreage, while many more have divided the land up among their sons. In Kenya women do not inherit land since they move to the plot of their husband. While the fact that women could not inherit land was outlawed in the 2010 new Kenyan constitution, overall, custom overcomes law and except in rare incidences the situation has not changed.

The small-scale farm is a family enterprise. Everyone in the family is part of the enterprise. During planting, I have seen the whole family participating. The small toddler was carrying a small tin can of seeds. The old grandma who does not have the strength to do anything else dispenses the water and food, while the stronger members, including the school-aged children, helped with the planting.

This is a picture from our kitchen door of our neighbor Ruth's plot. Ruth, a settler whose husband passed away a long time ago, has never sold any of her land. She lives in the building on the left, while her son built the house on the right.

Planting of maize and beans, which is best done as early as possible after sufficient rainfall, is usually finished within a week.

Frequently someone in the family has paid employment in Nairobi, other major cities, or elsewhere. Almost ten percent of the adult Kenyans work overseas and send home funds to help the family. This now is the largest foreign exchange earner for the Kenyan economy with almost half coming from Kenyans in the United States. Gladys did this for two years in Zambia, three in Pakistan, and twelve in the United States. Many times the family has spent considerable amounts of money for the education of this person who has paid employment. The family does not see this as a failure but as a success by the family.

People in Kenya retire when they are 55 to 60 years old which gives them many more years of active life. The custom is for the Kenyan with an income to build a house for the family in the rural area where he/she comes from so that he/she can retire there. Part of this income is used as investment in farming from buying cows to inputs such as seeds and fertilizer. Farming takes capital investments and this comes from the family members with paid employment. They see working in Nairobi or elsewhere as just a temporary sojourn. At the end of each month when people are paid, they overwhelm the transport system as they visit their original home. At Christmas time Nairobi is emptied of people as they return to their real homes in the countryside.

As Gladys and I are doing in Lumakanda, the retired Kenyans who have moved back to the countryside frequently take care of grandchildren with the employed family members helping to pay for their upkeep. This pattern is repeated from generation to generation.

Lumakanda, like most places in Kenya, has a large number of one room rentals. Here the people employed in town rent these cheap rooms for about 2000 ($20) per month so that they can support their family and invest in their homestead. If their home area is overcrowded, as is the case of the area where Gladys comes from, those who have the financial ability buy land elsewhere. Gladys decided to buy a plot here in Lumakanda because many of her relatives had moved into the area. A sister lives about three kilometers away (two miles), a late elder cousin lived right down the street, and another lives in nearby Kipkarren River where the local Quaker church is called Viyalo Friends Church after the name of the village where Gladys comes from.

These are one room rental apartments near our house. People who work in town, but live elsewhere, rent these rooms while at work, but travel to visit and invest back in the area where they originally came from.

Foreigners who come to Kenya are amazed at how warm and friendly Kenyans are, even if seemingly poor. I attribute this to the fact that Kenyans are well-rooted, secure in knowing their place in the world/life, and fiercely loyal to their land and extended family. Even if they have to leave the homestead to find employment or to go overseas to work, they invest in the area where they came from knowing that they will be welcomed to return when the time comes.

Mechanized or
Small-holder Agriculture?

Conclusion: A small-holder plot is four times as productive as a mechanized farm of the same size.

Whenever people talk about the world's population increasing to 9 billion or even more alarmingly 12 billion, the concern is how the world is going to feed all those people since already over one billion people are hungry every day. Inevitably the answer is more mechanization as done in the US with more fertilizer, more herbicides and insecticides, GMO crops, and other technological fixes. This is promoted even though there is already enough food grown in the world to feed even the largest figure; the problem is that too much rots because it cannot be gotten to markets, or it is turned into fuel, or wasted at the consumer level. What is ignored is that those one billion hungry people are hungry only because they are too poor to buy the food that is available.

To determine if small-holder farming or large mechanized farming here in western Kenya is more productive, I decided to calculate what Gladys grows on our small plot and compare it to nearby large farms. We hardly sell anything that we produce but eat it ourselves or feed it to our animals. Since we often just go out into the field and pick what we are cooking that day, there are no exact measurements, only estimates. I have tried to be conservative in my estimates. As to prices, which can fluctuate widely, I have used what it would cost us to buy the same item at the time of harvesting. We have three-quarters of an acre, but one-quarter acre is taken up with our house, lawn, driveway, outside kitchen, barn, chicken house, and woodshed. We therefore grow crops on half an acre.

Our biggest crop is corn (maize). We just harvested six bags of corn that are now drying in the sun. We have some corn still in the field, equivalent to another bag, and we harvested a lot of "green" fresh corn for roasting ("corn on the cob" as it is called in the US) in the last month before the harvest. We also gave the green stalks to the cows. The total is about eight 200-pound bags and at a value of $25 per bag this produces $200. We shelled the corn and the 8 bags of cobs will be used for firewood for a value of $8.

But in addition, we planted beans between the rows of corn which matures much faster, so we harvested the equivalent of a bag of beans, valued at $60. While the corn was drying in the field, we also planted some sweet potatoes and beans in between the rows. We can't plant too much sweet potato because there is no easy way to store the harvest, so we just dig it up when we want to eat it. The value of this second crop is $50.

We have planted Napier grass (also called elephant grass) to feed our cows, mostly along the boundaries. We cut the grass every two months and the cuttings cover seven days each time. At $3 value per cutting, this gives us $126 per year. Since this only adds up to 42 days in the year, we are planting more Napier grass.

There are three lots of collard greens (*Sukuma wiki*) in our plot. We use some of this for eating. Gladys cuts some for the chickens every few days. Since we have so much, a woman comes and buys $1 worth which she then resells in Lumakanda. At 50 cents per day, this equals $180 for the year. But this is only the beginning of the greens. There is black night shade (*manugu*) which is just a weed that grows in the field. Gladys once picked some of this in the US when we were strawberry picking. Another green, vine spinach (*nderema*) grows on the boundary hedges. Gladys just found some when I was quizzing her about the various greens for this posting. In addition, we grow spider plant (*sagaa*), amaranth (*terere*), slender leaf (*mitoo*), cowpeas (*kunde*), and jute mallow (*mrenda*). We also have three large pumpkin plants, but here we eat the pumpkin leaves, in addition to any pumpkins (which are what we call squash in the US), that might grow. Altogether, if we harvest some greens every other day of the year at 50 cents, this is $90.

We also have 14 cooking banana trees (plantains). These will give two big stalks per year at $5 each for a value of $140. In addition, we have 11 cypress trees, 40 grevillea trees (from Australia), and three indigenous trees for a total of 54 trees and we probably will plant more along our boundaries. Trees here grow 10 feet or more per year. After ten years, these trees may be worth $20 as firewood or $2 per year. Even though we only planted the first ones less than three years ago, we have already twice cut the lower branches so that the trees will grow straight and tall. We used the branches for firewood. Fifty-four trees at $2 per year equals $108.

Lastly, we have two sugar cane plants, one papaya tree, one avocado tree, and two mango trees. We can give this a conservative estimate of $25.

We can then total up the produce of our half an acre as follows:

Maize, 8 bags, @ $25 = $200
Maize cobs for firewood @ $1 per bag for 8 bags = $8
Beans, 1 bag @ $60 = $60
Second planting of beans/sweet potatoes = $50
Napier grass:
    6 cuttings per year for 7 days @ $3 each =$126
Collard greens @ 50c per a day = $180
Greens @ 50c per day for 180 days = $90
Cooking Bananas, 14 at 2 stalks each @ $5 = $140
Trees, 54 @ $2 per year = $108

Sugar cane, avocado tree, papaya tree, and
    2 mango trees = $25

This totals $987 for our half acre.

As we drive towards Eldoret every two weeks to go shopping, we pass many large, mechanized farms. Corn farmers in the US easily get 50 bags per acre, but here those farmers who farm well get only twenty 200-pound bags of maize per acre. At $25 per bag their income is only $500 per acre. There is no second crop as we do in our plot. This is only one quarter of the $1974 we would obtain from a full acre. Therefore our plot is four times as productive as a mechanized equivalent acre.

We have other major benefits to our method of growing. While the mechanized farmer uses diesel for plowing, harrowing, planting, and shelling their maize, we do everything by hand except for about 15 minutes needed to shell our six bags of maize. Then their maize needs to be shipped to the farmyard and subsequently to the place of sale. Our produce needs almost no packaging since we mostly pick it and put it directly in the pot. We cannot say that our food is organic because we put commercial fertilizer along with our cow and chicken manure on our plants. A few months ago we also had to spray the collard greens because they had bugs.

There is a further, frequently overlooked, aspect to the small-scale farming. If something happens climate-wise, we don't lose everything. A few years ago Gladys was planting some maize and beans together, but a hail storm came just as the beans were flowering and we got almost no harvest. The weather here can be finicky. In 2011 there was considerable drought in the corn growing regions, so harvests were substantially reduced. With our mixed farming, if one crop goes bad for one reason or another, there are still others which might grow well. We don't put all our hopes in one basket.

Seventy percent of the world's crops are produced by small-holder farmers. It is here, rather than the large mechanized farms, where the world can easily and substantially increase the amount of food grown.

Effort Needed for Small-scale Farming

These Reports from Kenya on small-scale farming are the results of what I have seen and learned over the years. These are actual descriptions of what farmers do, not what they might do or ought to do. This then emphasizes the strengths of their current activities. I also show what can be done better because some farmers are doing it already. Lastly I focus on how improvements do come about, emphasizing that, when something really works, it is quickly adopted through word of mouth.

I would like to start with a very famous small-scale farmer, Henry David Thoreau. The first year he cultivated about two acres and lived off what he produced. In 1845 he built a small house and lived at Walden Pond for a little over two years. The first year on his two acres he produced 4 bags of beans, six bags of potatoes, some peas and sweet corn. The second year he decided to cut the area to one-third acre. The first chapter of *Walden* or *Life in the Woods* gives exact details on this project. Here are some quotes from that chapter:

"I learned from my two years' experience that it would cost incredibly little trouble to obtain one's necessary food, even at this latitude (Massachusetts).

"For more than five years I maintained myself by the labor of my hands, and I found that, by working six weeks in a year, I could meet all the expenses of living. The whole of my winters as well as most of my summers, I had free and clear for study."

This field of maize is somewhat over an acre and plowed by oxen. Note the small garden at the right front. This is a small plot of cowpeas the leaves of which are cooked as a green. The large dark green leaves are pumpkin (squash) plants which in time will spread all over and produce more pumpkins than the family can eat.

Let me analyze what it would take to cultivate one acre of maize (corn) here in Lumakanda, Kenya. There are seven steps

- First cultivation
- Second cultivation
- Planting
- First weeding
- Second weeding
- Harvesting
- Drying and shelling

Each of these activities takes about 10 days of labor by one person. The total labor input for the year is 70 days, leaving the cultivator free for 295 days to do other activities. If done properly, the harvest would be twenty

200-pound bags of maize. This one person would consume one bag of maize during the year. The cost of inputs for hybrid seeds and fertilizer (this is not organic farming) would be the equivalent of about 3 bags of maize. This would leave a surplus of about 16 bags of maize for sale or to give away.

The hardest work by far is the first and second cultivation. Thoreau hired a team of horses to do this for him. Here people with an acre of maize might use two oxen to plow the field—this could be done in one day with two people and two oxen. Others like us hire a tractor at the cost of one bag of maize for each plowing. Therefore, the days of work have decreased to 50 days in a year and 14 bags of surplus.

Beans are planted between the maize and normally there would be one to two bags of beans per acre although this year (2018) due to excessive rain we are getting less than a bag per acre. Sometime people also plant pumpkins (called "squash" the US) in the same field. Unlike Massachusetts, there can also be a second planting on the same acre with beans, sweet potatoes, greens, or other items.

Many fields here are less than an acre. A person can plant one tenth acre and harvest two bags of maize (plus the other crops) which would cover the cost of inputs and seeds/fertilizer. A small plot like this would be done by hand so the total amount of time needed for the year's work would be 7 days!!!

In addition, this work is not repetitious and monotonous like being a toll booth operator in the U.S. I consider this one of the most boring jobs in the world as I calculate that the operator will take the funds from 10 people per minute, 600 per hour, 4800 per day for day after day after day.

This small plot of one-tenth of an acre (the size of a build-ing lot) next to our plot was cultivated solely by hand and will produce enough maize for one person for the year. The beans that were inter-planted with the maize have already been harvested. After the maize was harvested it was replanted with sweet potatoes and greens.

If a person is going to plant one acre of maize, he or she is not going to do it alone but with the whole family. Five members of a family might take two days of work-ing together to plant that acre. We have learned that, when it is time to plant, we cannot schedule a three-day workshop because everyone is in the field helping out. "Everyone" includes toddlers to the elderly – the toddler just holds a small tin can with a few seeds in it and sows a little here and there while the elderly person who no longer has the stamina for this kind of work distributes the seeds to the others, hands out drinking water, and oversees the tea break and lunch.

It is a total myth that the life of the small farmer is "nasty, brutish, and short". Rather this style of living leaves the person free for a lot of other activities, socialization, and, for Thoreau, studying and writing books. Therefore, do not be condescending to the small-scale farmer, but be envious. Then remember that the most popular hobby in the United States is gardening. "In spring 2017, the number of people who did gardening within the last 12 months amounted to 117.6 million in the United States." (from The Statistics Portal.) As any one of the gardeners will tell you, the food you grow yourself tastes so much better than that bought in the store.

What Small Farmers Can Do that Large Farmers Can't

There are many practices that a small-scale farmer can do that a large commercial farmer cannot. Most of these increase the productivity of the land while also providing for non-monetary insurance against crop failure. I will outline some of the major differences. Gladys and I live in the settlement scheme where around 20 -acre plots were given to Kenyans from the farms left by the British settlers at the time of independence in 1963.

As we drive to Eldoret 45 kilometers (30 miles) away there are large commercial farms sold to influential Kenyans at the same time. Therefore I am easily able to compare the two methods of farming.

1. Diversity of crops:

Large commercial farms — we are talking about hundreds to even thousands of acres — usually grow one or perhaps two crops. In our area the main crop is maize (corn) with some hay, wheat, and perhaps sugar cane. I have not seen a commercial farm in our area grow beans, cooking bananas, sweet potatoes, or vegetables.

Because we cultivate this small plot by hand, we can leave the trees. This photo shows our two navel orange trees (on left and center), two mango trees (in back on left center), one of two avocado trees (behind the orange tree in the middle), some of our banana trees (right side) with collard greens growing under them, and the tops of a row of grevillea trees along the boundary can be seen in the background on the right.

Most small-scale farmers grow a mixture of crops. As a result, they are not dependent on any one crop. When one crop does poorly, they can depend on an alternative crop. For example, because of the excess rain this growing season (2018), beans have not done well, but the maize has done very well. This diversity of crops leads to enhance food security for the small-scale farmer.

2. Inter-planting:
The practice in our area is to plant beans and maize

This is a large commercial farm growing maize on the road towards Eldoret. It is doing quite well this year (2018), but last year this field got no harvest whatsoever because the maize plants were stunted and did not grow properly. I suspect that the farmer was sold fake seeds rather than good hybrid seeds from a seed company. I felt sorry for this large loss that this farmer had last year. Note also that there are no trees in the field (only in the background where the ground slopes down) nor along the edge of the plot as is common with the small-scale farmer.

at the same time. Beans grow much more quickly than maize so that by the time the beans are ready for harvesting, the maize is just flowering and producing the cobs. Long ago research showed that this inter-planting gives a larger total harvest than planting just maize by itself. In addition some pumpkins (squash in the United States) can be planted in the field where both the leaves and the pumpkins can be eaten. Pumpkins take awhile to develop so that they cover the ground after the beans have been harvested and the maize is drying in the field. Large commercial farmers can't do this because the bean harvest is done by hand and cannot be mechanized. So they have to grow a mono-crop of maize. Note also that the beans, as a legume, put nitrogen into the soil that the maize can use as it grows.

This is our poor harvest of beans from one acre of maize, drying in the sun before the plants were beaten with a stick to make the beans separate from the pods. These beans had been inner-planted with the maize which was still growing in the field when we harvested the beans.

3. Double cropping:

Small-scale farmers frequently plant twice (and sometimes three times) per year, at least on part of their land. For example, as the maize is drying in one of our plots, we have cultivated between the maize and planted sweet potatoes and groundnuts. Beans and various greens are also frequently planted during the second short-rain cultivation. Clearly this increases the amount of food that can be harvested from the same piece of land. Large-scale farmers do not do this.

4. Animals:

Most small-scale farmers also keep animals. Chickens and cows are most common, followed by goats and sheep. A few farmers also have pigs, oxen (for plowing and pulling carts), donkeys, turkeys, geese, ducks, rabbits, and/or bees. Besides the meat, milk, and eggs for use or sale, the manure from any of these animals is spread over the crop land enhancing its fertility.

5. Mixing trees with crops:

Small-scale farmers frequently plant trees along their boundaries. Some also have a small wood lot, usually of eucalyptus trees. Fruit trees are often planted around the house. These include mango, avocado, guava, orange, pawpaw, and other traditional fruits whose English names I do not know. Most farmers also have some plantains (cooking bananas) and perhaps yellow eating bananas.

6. Erosion control:

Since for a small-scale farmer, land is scarce, a considerable amount of effort is put into anti-erosion measures to protect the soil. In addition to the style of ditch pictured on the next page, on some very hilly land, terraces are built with Napier grass of the edge to keep the soil from eroding.

This is the small road between our plot on the right and our neighbor, Ruth's, plot on the left. Note that both of us have planted grevillea trees on the boundary. These can be used after a few years for firewood or kept for 15 to 30 years when they can be cut for lumber for use or sale. If you look at the back of the picture you can see trees with rounded tops. These are eucalyptus trees that can be used for firewood, poles, and lumber. When we built our house we purchased one very large eucalyptus tree from a member of the Friends Meeting for $250 — this was enough to cut all the rafters we needed for the roof of the house.

This is an anti-erosion ditch through the center of our plot. When torrential rain falls and there is considerable run-off from the saturated ground, these ditches receive the excess water, allowing it to seep into the ground and thereby stops soil erosion.

Conclusion

These are just some of the more prominent methods that small-scale farmers use to produce much more per acre than a large commercial farmer. When I observe how much Gladys (with hired help) is able to produce from our half acre at home and two nearby acres of maize/beans, I realize how productive small farms can be. While I would assess that we are one of the better farmers in the area, there are many equivalent small-scale farmers near us and a few who are even more productive.

CHAPTER THREE: Physical Constraints to Farming

Three weeks ago we had a tremendous hail storm here in Lumakanda. Notice in the photo above of banana trees that some of the older leaves have been shredded by the hail with new leaves like the one on the bottom left are intact.

Farming in Kenya is extremely different from farming in the corn belt of the United States. Frequently Americans and Europeans don't understand the particular physical constraints here in Kenya and assume incorrectly that what is done in the northern hemisphere is not only the best way, but the way that others like Kenyans should follow and copy. I want to illustrate many of the constraints here in Kenya that do not exist in the US and Europe.

1. No snow: It doesn't snow here in Kenya. Dah, so what? In the US and Europe, it snows in the wintertime so that when the snow melts in the spring, the ground is saturated with moisture. This gives the plants a healthy start after planting. Here there is only the dry season and the wet season. When it finally starts raining at the end of the dry season, the ground is bone dry. How much rain should there be before a farmer plants? The answer is not clear. One of the rules of good farming here is "early planting" because, if one delays too long, the rains will have ended before the crop is mature. This year (2018) it started raining at the beginning of March when the rains usually start at the end of March. Some people planted during these early rains and, since the rains continued without a break, those early planters got a good harvest. But if the rains had stopped, they would have had a poor or even non-existent harvest.

2. Little topsoil: When I mentioned to an American that there was little to no topsoil here in Kenya, he wondered how people could farm at all. During both the rainy season and the dry season, insects including termites and bacteria continue to do their work of eating the vegetation. As a result there is very little organic matter in the soil. This means that the farmer has to increase the organic matter with manure and other organic materials and/or fertilizers.

3. *Drought:* Droughts are common here in Africa. Droughts also have a different meaning than in the United States. For example, a good number of years ago there was a drought in the US corn belt so that farmers got only 88% of their expected yield. Here in Kenya a drought can mean that one gets almost no yield or even, in severe cases, absolutely nothing, meaning that the seeds that were used for planting were wasted.

4. *Inconsistent rainfall:* There are two problems here. The first is that while the average rainfall seems adequate for a certain crop, the average is based on wild fluctuations. One year the rainfall might be half the average, leading to crop failure due to drought. Another year might be one and a half times the average rainfall, leading to flooding. This year (2018) with excessive rainfall in Lumakanda we received almost no yield from our beans which do not thrive in saturated soil. The "average" rainfall might occur only three or four times out of ten years. This has led to many disastrous projects promoted by commercial advocates from overseas.

The second problem is that the rain can be inconsistent. It might rain well for a certain amount of time but then it might stop for two, three, or more weeks so that the crops, even in a year with adequate rainfall, wilt during the dry spell leading to decreased crop yield. One year when I was in Tanzania, I went through one of these dry spells, but what was so antagonizing is that I could see the rain falling on the next hill, but time after time it missed the hill where I was living. The other hill had a good yield and my area did not.

5. *Hail and storms:* Fortunately there are no hurricanes or tornadoes. Nonetheless there are hail storms. Here is a 25 second video of hail in Kenya: <youtu.be/ObaNkcKLhPo>. We get a hail storm usually once or twice a year and they can be extremely destructive to crops. If it comes at the time when the beans are flowering, all the flowers are removed by the hail and no beans are produced. This happened at the refugee in camp where I was in 1964-65. We recently had a big hail storm here at our plot in Lumakanda and it shredded the leaves of big leaf crops. At other times high winds can blow over the tall maize stalks and torrential rains can lead to extensive erosion and mud slides which can be deadly.

6. *Weeds, insects and diseases:* Since there is no frozen time of the year here in Kenya like in the North America and Europe, during the dry season weeds continue to grow, insects multiply, and plant diseases are all ready to pounce on the newly planted crops.

The latest problem is that in the last few years army worm, imported from the United States, has invaded most of sub-Saharan Africa and is spreading rapidly, including here in Kenya. Army worms love maize, but will eat other crops. Farmers need to learn how to deal with the army worm and the plant geneticists need to develop resistant varieties of crops. There is always more to do in the battle between the farmers and their "enemies."

The small-scale farmer needs to know how to respond to all these threats to a bountiful harvest. There are many strategies to do this and the small-scale farmer must be well versed in the possibilities. Small-scale farming is an occupation that requires a great deal of knowledge and experience.

Constraints on Small-scale Farmers

Farmer in Kabaune village, Kenya, plowing his field with oxen. Photo: P. Casier / CGIAR via Flickr.

What are the constraints on small-scale farmers in Kenya from being more productive? Here are some of the main blocks to enhanced production.

1. Government neglect: The Kenyan government, like most governments throughout the world including the United States, supports the commercial, large-scale farmers at the expense of the small-scale famers. After the drought in 2017, in order to increase production of maize (corn), the government subsidized fertilizer so that a 50 kilogram (110 pounds) bag was sold for 1500/- ($15) instead of 3000/- ($30). The problem was that this subsidized fertilizer was only available to the commercial farmers. All small-scale farmers, including us, had to pay 3000/- ($30) for a bag of fertilizer.

2. Government keeps the price of food low: The government is based in Nairobi, the capital city of Kenya. Nairobi with 10 percent of the country's population generates 60 percent of its production. The urban population can easily put pressure on the government to try to keep the price of food as low as possible. Again, for example, during the drought last year the price of maize flour, the Kenyan staple food, increased dramatically. Since it was also an election year, the government subsidized the cost of the maize flour. This subsidized maize flour occurred predominately in the supermarkets in the main cities and towns, but often did not reach the countryside. This was supposed to terminate shortly after the August election, but when the election was annulled and there needed to be a second election, the subsidy was continued until after the second election.

A Kenyan small-scale farmer weeding her vegetables.

3. *Middlemen make the profits:* Shortly after independence in 1963, Kenya had well functioning marketing systems and cooperative societies. During the Moi years most of these collapsed to be replaced by middlemen who would buy the harvest as cheaply as possible and sell it at as high a price as possible. This is the current system. For example, a few years ago the Peace Centre on Mt. Elgon grew 3,000 cabbages to sell to support the centre. The first middleman offered 3/- (3 US cents) for each head of cabbage. The centre turned this down and the second offer from a middleman was 7/- per cabbage. The centre agreed and received 21,000/- ($210) for their cabbages. At this time cabbages were being sold at our local Kipkarren River market for 35/- ($.35) per head. Consequently of the selling price the centre received 20 percent of the value and the middleman received 80 percent. This is common across most of the crops sold in Kenya. The small-scale farmers do all the work, take all the risks, and receive the minimum for their efforts.

4. *Corruption:* Small-scale farmers also pay because of corruption. For example, many "fake" commercial farmers obtained exceedingly large quantities of the subsidized fertilizer discussed above. They had no intention of putting this on their non-existent farms. Rather they repackaged the 1500/- ($15) fertilizer in new bags and sold it to the small-scale farmers at 3000/- ($30) per bag, thereby making a gross profit of 100 percent. The taxpayers, including the small-scale farmers, are the ones who paid for this subsidy which was essentially stolen by the "fake" farmers. On a larger scale most of the cooperative societies and para-statal marketing organizations collapsed because of corruption. Coffee, pyrethrum, Kenya Cooperative Creameries (dairy), Kenya Farmers Cooperative, Kenya Meat Commission, numerous state-owned sugar factories, cashew and macadamia nuts, and so on, all collapsed to the great detriment to the small-scale farmer. In these cases commercial farmers have also been badly affected.

5. *Inefficiencies:* Due to terrible, poorly designed and maintained roads, small-scale farmers are sometimes unable to get their crops to market. In the rainy season — particularly this year when there has been excessive rainfall — roads become impassable. Milk and other perishable crops such as fruits and vegetables are particularly affected by these bad roads. Sometimes seeds or fertilizers are late in arriving and therefore are unavailable at the optimum time for planting/fertilizing. Farmers also frequently have to wait months and even a year or more to be paid for their crops. This is particularly a problem with sugar, maize, and coffee harvests. And when purchasing organizations collapse, the farmers suffer the loss.

A small-scale farmer watering his cabbages by hand.

6. *Lack of credit/crop insurance:* Farming takes costly inputs that are only recovered when the harvest is paid for. Just the seeds and fertilizer for an acre of maize costs about 8,000/- ($80) and in our area it is almost six months before there is any return. Since short term credit is on the whole unavailable, if the farmer is short on cash, he/she skimps on the inputs, thereby significantly lowering the yield. Moreover in some cases due to drought,

hail, and/or disease, the crop is a loss. There is no insurance to cover these types of losses so the farmer has to re-invest from other sources if he/she is able to do this. Selling a calf or cow is one possibility.

7. *Other issues:* The tyranny of having to pay school fees in January, April, and August at the beginning of each school term is a burden that forces farmers to sell crops at "throw-away prices," as it is called here in Kenya. Another possible problem is medical expenses if someone in the family gets sick. In other words small-scale farming is based on a precarious financial foundation.

To help cushion these slings and arrows of outrageous fortune, farmers either secure other jobs or engage in day labor. When we plant, weed, or harvest an acre of maize, work that requires up to ten people to do this in one day, we never have any difficulty in finding enough people willing to work for the day.

I have read that small-scale farmers in Kenya produce 70 to 75 percent of the crop production in the country. If the above constraints were properly addressed, how much more production would come from these small-scale farmers? Double the amount? Triple the amount? To state this in another way, only a fraction of the crop potential in Kenya is currently being used. Substantial increases are realistically possible that would feed the estimated increase of the Kenyan population in the next few decades.

Wastage after Harvest

It is estimated that one-third of the food harvested in the world is wasted before consumption. In some places in Africa up to one-half of the food harvested is lost before consumption. Some of the reasons for this include lack of proper storage facilities including refrigeration, improper storage leading to insect and mold damage, impassable roads, remoteness from markets, over-production of certain food items at harvest time, lack of adequate markets, corruption in the marketing of food, and so on.

Let me give a current example from Kenya. Charity Ngilu, the governor of one semi-arid county, visited India and secured an agreement from India to import green grams, also called mung beans, a drought resistant crop ideal for that county. The governor indicated that green grams would fetch 100/- ($1) per kilogram (2.2 pounds) which was more than the going price of 80/- ($0.80) per kilogram. She actively promoted the planting of the green grams as did some of the nearby counties with the same climate. With the good rains this year there was a bumper harvest. Then the Indian Government placed an embargo on the importation of green grams. Consequently there was no market and middlemen, if they can be found, are buying some of the crop at 30/- ($0.30) per kilogram. The farmers therefore were taking a huge loss as the Kenyan market was saturated.

Truck load of mangoes rotting when stuck on road to market.

Another example is that due to the heavy rains this year (2018) many roads, particularly, the local country roads, have become impassable. Farmers with perishable crops such as milk, vegetables, and ripe fruits were unable to get their crops to market. Likewise, the heavy rains led to the rotting of some crops in the field or during harvesting.

These losses are all from commercial, large-scale farmers who need to sell their harvest in the domestic or international market. This is where the post-harvest wastage occurs at the high rate of 33 to 50 percent.

Due to the excessive rainfall this season, part of our maize field became waterlogged and the maize did not produce cobs. Rather than let this go to waste as a commercial farmer would have to do, our workers, Mudavadi on the motorcycle, and Sidi, standing, cut the green stalks and brought them to our compound to feed to the cows.

My observation is that the small-scale farmer actually has a negative loss. What do I mean by this? The small-scale farmer can use parts of plants that large commercial farmers are unable to use so they just throw them away. Here are some examples. When sweet potatoes are harvested, the green vines are fed to livestock. When green maize (which is called "corn on the cob" in the United States) is harvested the still green stalks are again fed to the livestock. At the dry harvest of maize, the stalks are ground up for silage for cows and the cobs are used for firewood. Any maize that is damaged is separated and fed to the chickens. Greens never go bad on the way to market because they are picked daily from the garden as needed. If there is a surplus of an item that can't be sold, it is given away to relatives and friends.

The agricultural operation on our small farm is similar to that of everyone else in the local community. Gladys is the person in charge of the farm and knows how things are done here in Lumakanda. Small-scale farming is actually an occupation that needs a lot of knowledge and experience. Like our neighbors, we hardly ever waste anything, never anywhere like the 33 to 50 percent that the commercial farmer does. In addition, we have almost no transportation costs and little packaging costs. The major cost for packaging is $0.50 for the 200-pound bag to store maize. On the other side of the equation, if a crop does not do well (such as our beans this year because of the excessive rain), since the farm grows a great diversity of crops, this is not the great disaster that it is for the large farmer who is dependent upon large fields of one mono-crop.

Therefore, when comparing the small-scale farmer to the commercial large-scale farmer, one cannot just compare yields, but rather the "effective yield" which is the total harvest less wastage. By this metric the small-scale farmer produces much more consumable crops per acre than the large-scale farmer.

Chapter Four
Improvements for Small-scale Farmers

There are numerous methods by which Kenyan small-scale farmers can improve their productivity, sometimes dramatically. Obviously, removing the constraints that I mentioned previously would be extremely helpful. There are also many other activities that can be done. These can be divided into two categories, improvements that are developed by outsiders and improvements of the agricultural practices of the farmers themselves.

The most obvious example for external improvements is the development of better crops. These can include drought-resistant varieties, including those that mature more quickly. For example, Janet, my sister-in-law, just gave us six cuttings of a new variety of sweet potato that matures more quickly, by three months instead of up to six months. We have planted it to see how it does. We will then use the vines we have produced next growing season in a larger area. Other new varieties of plants can be resistant to disease or pests and/or more nutritious. There are a number of government-sponsored, non-profit organizations, and for-profit companies in Kenya that develop and promote improved agricultural crops. The latest need is to control the extremely destructive army worm that has just been imported from the United States and has spread over much of sub-Saharan Africa. Army worm-resistant varieties are in the process of being developed, but in the meantime maize and other grain harvests may be substantially decreased. We have not yet found any army worms in our maize fields.

Let me give three examples of innovations that we do on our plot.

A nice harvest of hybrid maize (corn). Notice that it is white. In the US corn is yellow. In 1963 President Kennedy sent US surplus corn during a famine and I was told that some Kenyans refused to eat the corn because it was yellow. In developing new and improved varieties of crops, one difficulty is when the new variety departs from what people are accustomed to. Currently the agricultural research stations are trying to make sweet potatoes more nutritious by adding carotene. This will make the traditional white sweet potatoes become yellow. Will Kenyans be willing to eat yellow sweet potatoes? Photo from Arthur Kihima's facebook page.

Page 60 (top) shows our farmer manager, Mudavadi, who has filled up a cloth sack with dirt that has one hundred holes in the sides where crops can be grown. This allows four to five times the output of plants on the ground. This is useful for people who have very small areas. I have seen this sack farming in Nairobi. But the sack costs 10,000 /- ($10). Rather than buy this bag that can be used over and over again, some people just cut holes in the sides of the normal 100-kilogram (220-pound) bag, used for storing maize and other grain. Page 60 (bottom) is that same bag about a month later with 100 swiss chard plants growing out of the holes.

The photo on page 61 shows a triple bag innovation that has recently become available. This inside bag can hold 90 kilograms (200 pounds) of maize or other grain. Notice the two plastic bags. The grain is put in the inner plastic bag and tied, then the second plastic bag is tied, and lastly the outside white bag is tied. This produces an airtight container so that moisture cannot enter the storage bag. This also keeps out oxygen so that any insects in the grain will suffocate. Consequently, insecticide does not need to be put into the bag as has been done traditionally. When we gave out surplus government food during the 2008 post-election violence, we found that young children, sick people, and the elderly could not eat the stored grain because the insecticide gave them diarrhea. So the lack of insecticide is clearly a great health benefit. This bag, obtained here in Lumakanda, at the local agricultural/livestock store, cost 280/- ($2.80), but can be used for at least five years. The usual bags cost 50/- ($0.50) plus the cost of the insecticide.

Triple bag system that corn to be stored longer by avoiding effects of moisture and insects.

Traditional three-stone fireplace

Newly developed more efficient wood stove

The top photo on page 62 is the traditional three-stone fireplace inside most kitchens in rural areas. Because of the smoke these are placed in a separate building outside of the main house. Our outside kitchen has a chimney which many traditional kitchens do not, but nonetheless the room is frequently quite smoky.

The bottom photo on page 62 is a newly developed more efficient wood stove. It is advertized as using half the firewood and giving one-third the smoke as the traditional three-stone method. One of the promotions on the side of the box said that the stove was invented in America. It is sold in the supermarkets in Eldoret for 3,890/- ($38.90). We also bought a similar more efficient charcoal stove. Loreen, our niece, said that when she is cooking chapatti, she prefers this new stove to the traditional charcoal stove because, since the fire is confined to the pot on top of the burner, she does not sweat from the wasted heat given off by the traditional stove.

These are just three examples of the many, many innovations that have been developed. I am certain that many more will also be developed to increase the productivity and decrease wastage for the small-scale farmer.

Enhancing Productivity of Small-scale Farmers

Joseph Khisa prepares his family's land for for planting in Chwele, Kenya. Notice the use of a string to make sure that the row is straight. This is a recommendation of the One Acre Fund. Also if you look carefully at the string you can see that there are small pieces of cloth that show where each seed should be planted. This insures that proper spacing is used in the planting so that seeds are not wasted or sowed too far apart.

Kenyan small-scale farmers can significantly improve their agricultural practices. When I was in Kenya in the 1960s, shortly after independence, Kenyan farmers were enthusiastic in adopting new techniques – hybrid maize, grade cows, growing coffee, tea, and other cash crops, and so on. They were advised by well-trained agriculture and livestock government officers. It was an exciting time. Unfortunately, much of this dissipated during the Moi years. Farmer training centers were closed and the agricultural extensions officers were curtailed and abolished, which are among the many factors that have impeded improvement in Kenyan agriculture.

Many non-government organizations promote agricultural improvements. Here I will cover one of these, the One Acre Fund. I pick this program because three of my five sisters-in-law are enthusiastic members of the One Acre Fund, and a fourth was a participant until she got sick (but says she plans on rejoining). The fifth sister-in-law was once a member and resigned because she didn't like it.

In 2017 the One Acre Fund, which began in Kenya in 2006, served 238,000 farmers in Kenya. This is what its website explains <oneacrefund.org>:

"We offer a complete bundle of services, using a market-based model that helps our organization remain financially sustainable and expand to reach more and more farmers every year. Here's how our model works:

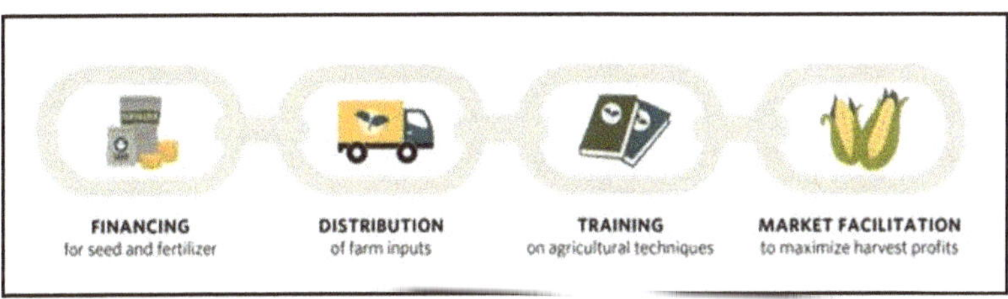

"**Asset-Based Loans.** Farmers receive high-quality seeds and fertilizer on credit, and we offer a flexible repayment system that allows them to pay back their loans in any amount throughout the loan term.

"**Delivery.** We deliver inputs to locations within walking distance of every farmer we serve.

"**Training.** Farmers receive training throughout the season on modern agricultural techniques.

"**Market Facilitation.** We offer crop storage solutions and teach farmers about market fluctuations, so that they can time crop sales to maximize profits."

Since the One Acre Fund can purchase seeds and fertilizer in bulk from the suppliers, the problem of fake seeds and fertilizer is solved. Moreover, even if the farmer is short on cash for one reason or another, their loan fund means that the inputs are available at the appropriate time. One year due to a shortage of hybrid maize seeds we had great difficulty finding the variety we wanted to plant. Although they don't mention this on the website, as soon as a new crop variety is available, the One Acre Fund is able to quickly introduce it to the farmers. They also take soil samples from farmers for testing in their lab in Kakamega and, if necessary, recommend that the farmers spread lime which is often deficient in the soil. The new plastic liner storage bags that don't need insecticide that I mentioned above are also promoted by the One Acre Fund.

The One Acre Fund also promotes the planting of grevillea trees which have deep roots and do not dry out the soil as the commonly planted eucalyptus trees do. Their participants have planted eight million of these trees in Kenya. We have about forty on the boundary of our plot. One Acre also provides solar lights, vegetable seeds, cook stoves (I don't know if it is the one we have that I have described earlier), and sanitary napkins.

The One Acre Fund can also play an important role that is not mentioned in their information. The agriculture research stations can develop an improved crop, but then they need to have the commercial seed companies propagate the seeds in substantial amounts. The seed companies are reluctant to do this unless they can be assured of a large market. The One Acre Fund can provide this for them since, if they decide to promote a new variety, they will have enough need for substantial amounts, thus encouraging the seed companies to produce the new variety.

There is not one magic bullet that is going to solve the increase in production by small-scale farmers. Rather there are going to be innumerable small steps, many which will be intertwined with other improvements. While the possibilities are many, the important aspect is that the small-scale farmers are willing to adopt them. If they don't adopt a particular option, they have a good reason which, if understood, should be accommodated.

Animals

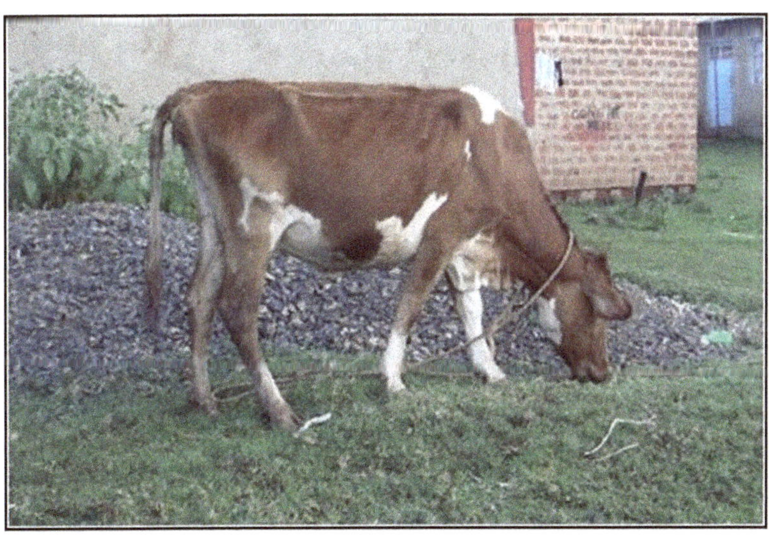

This cow is owned by one of our neighbors. Notice that the cow is tied up and is eating grass along the roadside. This is a very common practice. Unfortunately this does not allow the cow to produce much milk. We do what is called "zero grazing" in that we keep the cows in the barn and bring them their food, mostly chopped up Napier (elephant) grass. This gives considerably more milk production.

Almost all small-scale farmers in Kenya keep animals with most keeping a variety. Here is the list according to the 2009 census when the Kenyan human population was 39,000,000 million:

Exotic cattle (European breeds) = 3,355,407

Indigenous cattle (zebu cows) = 14,112,367

Sheep = 17,129,606

Goats = 27,740,153

Camels = 2,971,111

Chickens (2016 estimate) = 31,000,000

In addition to cows, some farmers also keep other animals — goats, sheep, camels (in the arid and semi-arid parts of Kenya), donkeys, pigs, turkeys, ducks, geese, rabbits, dogs (which live outside and guard the house), and cats. This diversity of animals helps support the small-scale farmer and is a great enhancement to his/her quality of life.

Note that goats are the most common livestock in Kenya as its meat is the favorite meat in Kenya. The favorite dish is *nyama choma*, roasted goat meat. I don't understand why Americans do not eat goat which is the leanest, healthest of all meat and goats eat brush and don't eat grain or grass. Americans consume the most meat of anywhere in the world, but their consumption of goat is one-quarter pound per year and I expect that most of this is eaten by people from Africa and the Caribbean living in the US.

I will focus on the two most important animals, cows and chickens.

Cattle: Cattle serve four purposes in the life of the small-scale farmer. First the cows give milk which Kenyans, following the British custom, pour in their tea. Not only does this taste good, but it is very healthy for children.

The second use is for meat. Cows in Kenya are all grass-feed as they are not given grain to eat like in the US. Consequently the meat is healthy and not fatty, nor do the cows consume food that can be eaten by humans. Here in Lumakanda we pay 320/- for a kilogram of meat ($1.45 per pound). There are four butcher shops in Lumakanda that sell beef. The slaughter house is down the road in Gladys' cousin's plot and one or more cows are slaughtered each day. The person who kills the cows in the kosher way by slitting the throat is always a Muslim. Since Muslins do not eat pork, the butcher shops do not sell meat from pigs. There is a special butcher shop that sells pork and I assume that the slaughtering is done elsewhere by a non-Muslim

The third is for their manure which is spread on the fields to enhance fertility.

The fourth, and often neglected purpose for keeping cows, is pecuniary. It is not surprising that the Latin root of this word means "cattle." People keep cows as a easily accessible savings account. If there is a need for a large expense such as school fees, health problems, building a house, and so on, a cow can easily be sold to obtain the needed funds. Every Thursday in Kipkarren River near our house there is a cattle/goat/sheep market where anyone can sell their animals. We have used it on occasion. Moreover, cows give a much higher return on the savings account. We are receiving 0.05 percent interest on our saving account in the US and 1.5 percent on our savings account in Kenya. A cow will give a calf about once per year and three years after a heifer is born, she can give birth to another cow. A nice full grown cow will fetch from 30,000/- ($300) to 75,000/- ($750). Those Maasai so beloved by tourists who are walking around with 100 cows are grazing at least 3,000,000/- ($30,000) worth of livestock.

Except for bulls who have been castrated to become oxen, there are very few bulls – a female calf is a birth of joy. The reason is that here in highlands of Kenya most farmers use artificial insemination with the exotic breed. This means that over time the pure traditional zebu cow become more of the exotic breed—one half, then three quarters, then seven eights and so on. Since the exotic breeds give much more milk, over time the production of milk will continue to increase, no doubt keeping up with the human population increase.

Many farmers just let the hens run around the yard searching for food, but they return home at night to roost. This is one of our hens with her chicks that we allow outside the chicken house where we keep the adult hens and one rooster.

Chickens: The chicken or the egg. It is not which came first, but rather, when a hen lays an egg, do you eat the egg or let the hen sit on it to hatch a chick? Most small-scale farmers in Kenya keep chickens. We usually have between twenty and thirty hens and chicks with usually one rooster. If there were more roosters, they would fight too much. Cock-fighting, by the way, is not a Kenyan custom. We do let them out sometimes in the late afternoon to search for bugs and eat some grass which gives their yolks the dark yellow color. Many

70

small-scale farmers who keep chickens just let them run around and scratch for food.

Like cows, chickens have four purposes:

The first is the eggs, while the second is meat. All roosters except one are slaughtered at home for dinner. Unlike beef which is cheap, chickens are expensive. One adult hen/rooster will cost between 500/- ($5.00) and 750/- ($7.50). Just before Christmas when everyone wants a nice chicken for their Christmas day festivities, the price rises to 1000/- ($10.00) or more. When chickens are kept in a chicken house like we do, the manure is spread on the fields.

The fourth purpose of chickens is most interesting. A chicken is the major gift that people give each other. When Gladys' father was still alive and we visited him, we would often give him a chicken for a present. His custom was to slaughter a chicken to honor our visit, but it could not be the chicken we brought. When relatives come to visit us, they sometimes bring a chicken. Children are also given chickens as presents. Under the custom here the chicken is then the "property" of the child. For example, grandson Brian was given a nice rooster, but we could not slaughter it until Douglas, his dad, came to visit us. This giving and receiving of chickens is a symbolic sign of relationship and friendship.

Some people also keep turkeys, geese, ducks, guinea fowls, doves, or rabbits. Rabbits aren't in the same class as a bird, but their meat is similar to chicken. The advantage of rabbits is that they eat grass and greens while chickens eat maize.

Hawks are major predators of chickens. Snakes are also predators which is the main reason that Kenyans kill any snakes that they can. In a bizarre case with one of our neighbors' chickens, the mother hen scratched a bees' nest and the bee swarm stung to death the hen and all her chicks. Greater fatality of domestic birds is caused by disease. Many or all of the fowl can quickly die if the appropriate medicine is not given to them.

A neighbor's goat tied up to eat the grass and shrubs plus a free range hen

There are two categories of chicken in Kenya. The most common in the rural areas are the local chickens called "kienyeji chickens", while others breed exotic varieties from the US and Europe. We keep kienyeji chickens. People prefer the kienyeji eggs because they usually have harder shells and yellower yokes.

The advantages of the exotic breeds are that they grow faster and lay more eggs. Their disadvantages are that they must be fed chicken mash, are more susceptible to disease, and must be kept in enclosed chicken coops. Most Kenyans think that kienyeji chickens taste much better. I agree that the industrial chicken is bland.

No wonder Colonel Sanders of KFC must use all those secret spices to make its chicken edible. There are at least nineteen and counting KFCs in Kenya. Gladys and I ate lunch at the recently opened outlet in Kenya. Their chicken here is as greasy and expensive as it is in the United States.

A battery cage for laying eggs as used by commercial poultry keepers in Kenya. Notice that the eggs are brown.

Battery cages for laying eggs, which are banned in the European Union, California, and other places, are vigorously promoted here in Kenya. All eggs in Kenya are brown, not white as in the US; there is no difference between brown and white eggs.

Coffee

A farmer tends to her coffee plants. (Nation Media Group)

A discussion on coffee growing in Kenya is useful to understand the problems of small-scale farmers. Coffee is not grown near Lumakanda, but it used to be grown in Gladys' home area near Chavakali and it is still grown on the slopes of Mt. Elgon which we can see from our kitchen window.

Let me start with an anecdote. Around 2004 I was in Burundi and bought a one-half kilogram (1.1 pounds) of coffee for the equivalent of $1.50. This coffee was grown, roasted, ground, and packaged in Burundi. This was a present for my aunt in the U.S.

When I gave it to her, she wanted to pay me $9.00 for it since this was the price of an equivalent amount of coffee in the U.S. In other words the coffee was six times the price in the US, meaning that the middlemen, importers, US coffee companies and others were making 84 percent of the money, leaving the Africans with a small fraction of it.

Today 70 to 75 percent of the coffee produced in Kenya is grown by about 700,000 small-scale farmers, while the rest is grown on large estates. These farmers have an average area of one-half acre under coffee. One of the major reasons that these farmers like to grow coffee is that the return from a small plot is much higher than could be obtained from any other crop. Note that Kenyans do not drink much coffee so that 93 percent of the coffee grown is exported for the international market.

At one time coffee was Kenyan's largest foreign exchange earner. Production peaked in 1989 at 130,000 tonnes. It is now dropped to around 40,000 tons. As a foreign-exchange earner, coffee is now behind tea, horticulture, tourism, and remittances from abroad.

Why this decline? It can be traced to external and internal causes.

External causes: While world consumption is fairly static, growing at only about one percent per year, production can vary significantly, often due to frost in Brazil's coffee growing region. In such cases the price of coffee beans increases and the farmers receive a decent return. But in other years of high coffee production the price decreases and the return to the small farmer barely makes the growing of coffee worthwhile. These fluctuations are discouraging to the farmer. Coffee takes two years after planting to begin producing and maximum production is two or three years after that. The grower is therefore dependent on the vicissitudes of the world market. When the price collapses, some farmers neglect their coffee or even uproot it.

Yet another problem is that the international community, including the World Bank and International Monetary Fund, pushes every coffee growing country to increase its production to earn more foreign exchange. This leads to overproduction of coffee and a decrease in price. In order to earn more foreign exchange, the Kenyan government is again encouraging farmers to plant more coffee.

Another issue is that the government can and does tax coffee exports, which takes some of what could have been given to the farmer to support the government itself. It is impossible for the government to tax the maize, beans, and collard greens like we grow in our farm, so it is in the interest of the government to promote export crops at the expense of local food production.

Internal reasons: All these small-scale coffee farmers were organized into coffee cooperatives. At the time of independence, these were strong institutions giving benefits to their members. But over the years mismanagement, nepotism, and corruption entered into the coffee societies. Many of them collapsed. Even those that continued often did not pay for a year or more after the coffee beans were delivered. As a result, farmers were forced by the need for funds to sell their berries to middlemen at poor prices. Recently in the news there have been reports of robbers breaking into coffee warehouses and stealing the berries, resulting in further losses for the cooperative societies and their growers.

As a result of both these external and internal causes, many small-scale coffee farmers have uprooted their coffee for more lucrative crops or, in some cases, particularly north of Nairobi, housing developments.

The government has been trying to support more coffee farming. For example, in Bungoma County next to us, the county government is giving away free coffee seedlings to farmers. Without broad improvements to the whole system, I am not sure if this will be effective.

There are some possibilities. For example, Java House, "Africa's favorite Coffee-led Restaurant", as it advertizes itself, is a coffee company based in Kenya that is similar to Starbucks in the US. It has 51 outlets targeting the high end market and sales last year of 4 billion shillings ($40 million). It uses local Kenyan beans that are roasted in its stores. Will this lead to higher coffee consumption in Kenya and a steadier market for its coffee beans?

One of the major issues is that the big coffee companies control the coffee trade. Most agricultural products from East Africa are sold without processing, while the exporters control the processing process and thereby reap the added value. There are attempts by African countries to add value to their coffee by processing it before selling it internationally. Rwanda Farmers Coffee Company owned by the six coffee cooperative societies in Rwanda has set up its own roasting and packaging line called Gorilla Coffee to export processed/packaged coffee. This year it has just sent 10 tons of Gorilla Coffee to the United States. Note from the picture that this coffee is labeled as "beyond fair trade".

While this is the story of coffee in Kenya, similar stories can be told about pyrethrum (a natural insecticide), macadamia and cashew nuts, mangoes, passion fruits, and many other crops. If all of these were properly organized, Kenyan small-scale farmers would be much more prosperous.

Fruits and vegetables available in one of the kiosks in Lumakanda.

The Local Market

The purpose of commercial farming is to sell the harvest to outsiders to secure income and profit. An additional purpose for export crops, such as, coffee, tea, and cut flowers, is to earn foreign exchange for the country. Government officials, economists, and development experts love export crops and promote them vigorously. But the goal of small-scale farmers is to provide food for themselves and their neighbors. This Report from Kenya discusses the local markets in Lumakanda town.

Lumakanda is the government seat of Lugari sub-county, so there are government offices, two police stations, schools, and a hospital, with many supporting businesses. As a result, there are many employees who need to buy food. Moreover no small-scale farmer is self-sufficient as they will always need to buy some items that they don't grow or don't have at a particular time.

One of the produce stands in Lumakanda

One day I walked around town and counted twenty-three places that were selling fruits and vegetables. Moreover, there were stands that were not open during my count and I probably missed some since I didn't cover every street in the town. All of these are owned by women. I also counted six women who just laid out their goods on some cloth on the ground. I once read in the newspaper that sixty percent of Kenyan women have a side occupation.

The most ambitious of these kiosks are major small businesses that are open all day. Women go to Kipkarren River, five kilometers (three miles) away, on motorcycles to buy what they need at the market on Thursdays and Sundays. They also bring produce that their family and neighbors produce. Whenever we have an excess of any crop we have no difficulty selling it wholesale to these market women.

A produce stand attached to a shop

Other smaller kiosks open only from about 4 to 7 PM. Some are connected with the regular shops in town as an outlet selling soap, sugar, cooking oil, and other common consumer items sold in the shop.

The vegetables offered by women who just put their goods on a cloth.

Almost all the stands sell tomatoes and onions, used by everyone in the local cuisine. Greens of all kinds including cabbage and collard greens are also common. Then there are other less common items such as cooking bananas, omena (a very small fish from Lake Victoria that is eaten with ugali), carrots, butternut squash, and various fruits including yellow bananas, oranges, and mangoes. In addition to produce, some of the women sell charcoal or firewood.

The women who sell from a cloth on the ground specialize in the various greens that they probably grow themselves. When they don't have anything to sell, they don't come to town. There are two different boys who come to our house on some afternoons selling greens to us and their other customers. I am sure these were grown by their mothers. Two women also sit outside the Lumakanda Township School selling small pieces of sugar cane to the students as they leave school late in the afternoon.

Milk sold from the afternoon milking at the kiosk shown in the previous picture.

Milk is also sold in town. There is one man who sits at the main corner in town in the morning selling his milk. There is also a "milk bar" in town which is a regular shop. This is all local milk. When we have a surplus from our afternoon milking, we sell it to one of the main merchants in town who resells it.

There are four butcher shops in town that sell beef and one that sells pork. People slaughter chickens themselves. If a person wants to buy a chicken, if he/she can't get one from neighbors, he/she has to go to Kipkarren River to purchase it.

Lumakanda is in no way unique. These produce kiosks are everywhere along the roads. I have no doubt that these activities are profitable in time and money for the women who do them. If they weren't, they would stop. Note that people do not have refrigerators so they need to buy produce every day or two. The result is a thriving local economy as money continues to circulate in the community. One piece of evidence for the circulation of money is that the 50/- (50 US cents) and 100/- ($1.00) notes are all crumpled, dirty, and torn from continued use. Economists don't like this local economy because this trade cannot be taxed. I suspect that it is not even adequately included in the gross domestic product of the country. As the small-scale farmer sells and buys from his/her neighbors, this becomes another method of building community and resiliency for the farmer.

Machakos District: What Works

Iveti Hills, 1937 (Colin Maher, February 1937) From 1968 to 1970 I could see this view from the Mua Hills Secondary School where I was the principal.

"In 1937 some early efforts of the Forestry Department in afforesting the summits could be seen above the largely treeless lower slopes. On these slopes Barnes [the photographer] remarked that 'numerous small gullies can be seen starting out of abandoned or closed native shambas (fields).' The rills and gullies fed unchecked into the scarred and rocky stream channels. Settlements were sparse. However, a start had been made in planting hedges and woodlots.

"The Machakos Reserve is an appalling example of a large area of land which has been subjected to unco-ordinated and practically uncontrolled development by natives whose multiplication and the increase of whose stock has been permitted, free from the checks of war and largely from those of disease, under the benevolent British rule.

"Every phase of misuse of land is vividly and poignantly displaced in this Reserve, the inhabitants of which are rapidly drifting to a state of hopeless and miserable poverty and their land to a parching desert of rocks, stones, and sand." (Colin Maher, February 1937).

As indicated by the above quote from 1937, the semi-arid Machakos District (now Machakos and Makueni Counties) just east of Nairobi, was considered a hopeless disaster by the British. This quote is taken from a unique study of the agricultural development of the district from 1930 to 1990. It is published in a book titled *More People, Less Erosion: Environmental Recovery in Kenya* by Mary Tiffen, Michale Mortimore, and Francis Gichuki (1994).

This is where I lived in 1968 through 1970 so I was very interested in the research. I bought the book when it came out but gave my copy away. When I started writing this series I thought I should re-read this book. I looked it up on Amazon to see if I could get a used copy and found one available for over $500. This was out of my price range. Then recently I looked again and saw one available for $22. I bought the book and re-read it.

How similar is this 1937 prediction of doom to those current doomsayers who pronounce massive food shortages by 2050 or 2100 leading to land degradation and starvation. These predictions ignore the capabilities and resiliency of small-scale farmers to produce the necessary food increases.

Picture of these same Iveti Hills in 1991. "Trees and houses are now ubiquitous, and only one substantial area of uncultivated land survives. Apart from this area, the gullies have become revegetated. All cultivated land is terraced. Among the trees, the range in size suggests they are regenerating."

The Machakos District, populated by the Kamba tribe, is prone to drought. There are two growing seasons in a year. In the 96 years between 1894 and 1990 there were 90 droughts during the 192 growing seasons. Consequently water use and conservation are critical in producing adequate food for the population. During the 60 years from 1930 to 1990, with a three percent annual population increase, the population of the district increased six-fold. Nonetheless there was a threefold increase in output per person and a ten-fold increase in the value of output per hectare (2.5 acres). Therefore people in 1990 were much better off than they were in 1930.

I will leave out what the government and NGO experts advised that failed. Rather I will stress what worked. Many innovations were tried during this period. The farmers got their ideas from themselves, from visiting other parts of the country and even the world, and possibilities promoted by the agricultural officers. In many cases an innovation started in that part of Machakos District which was close to Nairobi. This meant good roads for transport of crops and therefore the ability to receive good prices for whatever was sold. When an innovation did not work out, it was discarded.

A current picture of the Mua Hills. Notice the terraced fields on the right side and the artificial pond in the foreground. Since this is a semi-arid region with frequent droughts, all possible measures are needed to conserve water.

If it did improve production, it was quickly adopted locally and within a few years had spread throughout the district. The farmers were not conservative and resistant to change. Rather they wanted proof that an innovation worked. Improvements were mostly propagated by word of mouth.

For example, take the issue of terracing. Since all water had to be conserved in order for the crops to grow successfully, particularly in drought years, it was necessary to terrace all fields. There are different methods of terracing. The Kamba farmers quickly realized that the most labor intensive bench terracing was the most beneficial. In this case a sloping field was made flat with a backwards slope so that all rain water soaked into the ground.

This is a nice picture of bench terraces. Note the well-established grass strips along the bunds (terraces), the fruit trees planted in the field, and the banana trees at the back of the top benches. These were all innovations adopted by the farmers.

Making these terraces is a lot of work. My first father-in-law, Wilson Malinda, would go out every day about 5 PM and throw dirt up for an hour as he slowly made the bench terraces on his hilly plot on the Mua Hills. Philip Nsioka, the chief of the Mua Hills, once took me to his plot in nearby Mitaboni. He had built bench terraces on steep slopes which were about five feet tall. He told me that a small stream at the side of his plot became dry during the dry season but after he built the terraces, water ran in the creek all year around. His harvest was immense as he had trucks coming from Nairobi to buy his tomatoes and other produce. This impressive farm was one of the reasons the government had appointed him chief of our area.

I could continue listing many other improvements, but the lesson is clear. Small-scale farmers can dramatically improve their agricultural production. Another lesson from Tiffen, *et al,*1994, is that the farmers will

This is a picture of my son, Tommy, learning to plow with oxen on his mother's plot in the Mua Hills. (He, Joy, and grand-daughter Jayla visited Kenya for a week in November of 2018). He is cutting the furrows to sow beans when the short rains arrive. One of the major innovations that the Kamba people of Machakos adopted on their own was the use of oxen, plows, and ox wagons. The vast majority of the Kamba farmers own a plow and two oxen. Those without them rent the oxen and plow from their neighbors. The people in Rwanda and Burundi have never adopted oxen and plows. All small-scale farming is done by hand. I could never understand why.

quickly adopt those things that work well and will as quickly abandon those that are not useful. Since many of these improvements that were adopted were not recommendations from the government and NGO experts, the book concludes with the sentence, "It is the latter [experts] who need to learn the virtues of humility." (Mary Tiffen, Michale Mortimore, and Francis Gichuki (1994), page 285.)

Land

Land. That is the big issue. To be a small-scale farmer, a person needs to have access to land. The most common method of getting land is by inheritance from the father. Only sons inherit land as the daughters move to their husband's land. How much land a man receives as his inheritance has a lot of variables including the number of wives and sons his grandfather and father had, the original amount of land the ancestor had in the year 1900, and the many vagaries of life over generations. For example, I know a person whose father had 84 acres. But he had four wives and seventeen sons so each of them got about 5 acres. Then that person had four sons, so each one had only a little over an acre. To compensate for this small amount, he bought land elsewhere, but his sons preferred to remain in the home area.

The concept that a certain person owns a piece of land was a British importation into Kenya. This private ownership of land with a title deed is the major method of land holding in the former white highlands. Nonetheless there are major problems due to corruption in the land offices where land title deeds are given for bribes and fake title deeds are produced by the "land cartels" that specialize in this fraud. The "owners" of these corrupt transactions quickly sell "their land" to unsuspecting people. The Kenyan courts are full of land cases over disputed title deeds.

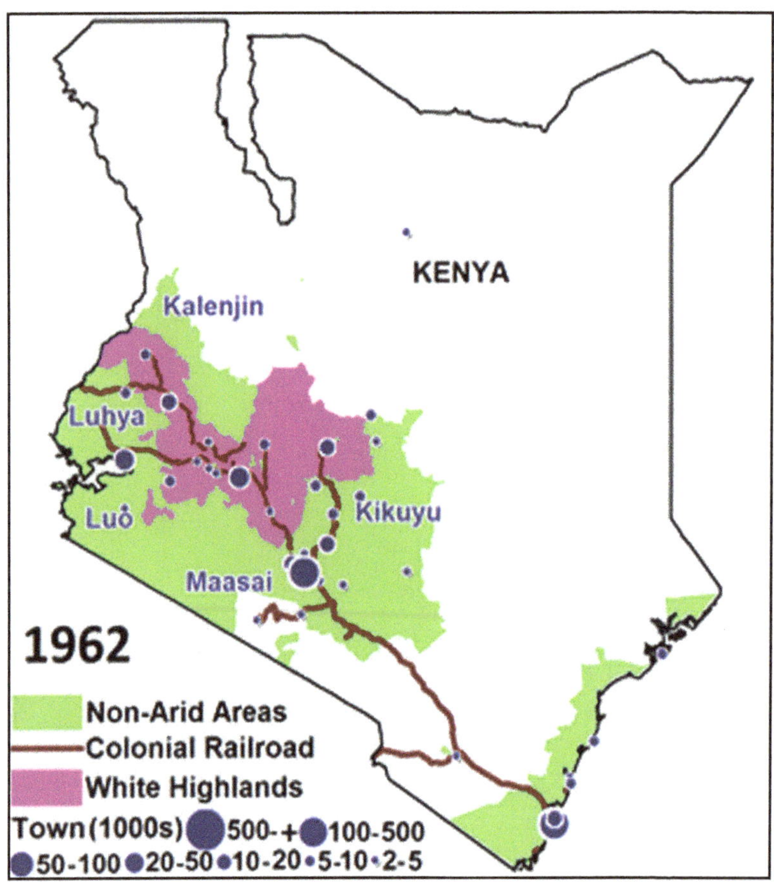

A map of Kenya just before independence showing the 20 percent of the country that is arable land and the significant portion controlled by the British white settlers, usually the best of that arable land.

In the African areas, the situation is much more fluid and complex. When the British arrived in Kenya, there was a low population density and no large commercial farms. Consequently there was a lot of available land and each person, usually with the agreement of the surrounding relatives (the "elders"), a man could obtain as much land has he could properly use. If his needs declined, then the land returned to the community and could be allocated to someone else.

Another issue is that before British colonialism, people and whole groups of people moved because of drought, sickness, or fighting. The different tribes in Kenya were frequently intermixed. When the British came, one of their strategies was "divide and rule." To implement this, they drew boundaries on the land and everyone within a certain boundary had to be of one specific tribe. For example, in the Machakos district, on one side of a line were Kambas, on the other side of the line, were Maasais.To this day there are disputes, sometimes violent, over these tribal boundaries since both sides can usually claim that at one time they inhabited the same area.

The Kenyan Independent Electoral and Boundary Commission not only conducts elections, but decides the boundaries of the various counties and constituencies. In Kenya this is based on ethnic considerations -- first as tribal, then as sub-tribal, then as clans, and lastly extended families at the local level. Politicians naturally try to manipulate the system to their benefit, similar to gerrymanding in the United States, but on ethnic, tribal lines. People who are not from the dominant tribe or ethnic group are considered "forcigners". When elections come around, aspirants will say things like, "The fields must be weeded," meaning the "foreigners" need to be expelled. This leads to the election violence that is common in most Kenyan elections.

As the population increased and the British took half the arable land for European settlement, land available for the small-scale farmer became scarcer. By the time of independence in 1963, most of the arable land was allocated so that there was no more vacant land for people to move to. The Million Acre Settlement Scheme in the former white highlands was one attempt to successfully allocate more land to African small-scale farmers.

Another method of obtaining land is to buy it. Some of the larger former European farms have been divided up into smaller lots. At independence land was quite cheap but most people did not have the available cash to buy land. Now land is quite expensive. A friend of mine who worked in agriculture in Iowa emailed me recently that arable land in Iowa is around $7,000 per acre. Agricultural land around Lumakanda is now $8,000 per acre and increasing every year as population increases but the land area does not. At this expensive rate, only the wealthy are able to buy land. Usually this is in small lots of an acre, more or less. In the over-populated area of Chavakali where Gladys originally came from, land is not sold and, if it were, it would be extremely expensive. The landless who live by day labor do not have the financial ability to buy even a small plot of land.

As the average small-scale farm has become smaller, the result has been intensifying the agricultural productivity of the land. This is a continuing process. As one can observe the differences between the best small-scale farmers and the worst ones, there is a large possibility of improvement. Moreover,

- if the government allocates sufficient resources to the agricultural sector,
- if agricultural extension for the small-scale farmer is adequately funded and promoted,
- if corruption is weeded out of marketing,
- if farmer organizations and cooperatives are promoted at the expense of middlemen,
- if more productive, drought resistant new crop varieties are developed and disseminated,
- if crop prices are set at beneficial rates for the farmer,
- if rural roads are adequately built and maintained,

- if unused and minimally used areas of large-scale farmers are turned over to small-scale farmers, ... then small-scale farmers can more than feed Kenyan's growing population.

Landless People

If you have read this far, I hope you have been convinced of the benefits small-scale farming has for the people in Kenya. Unfortunately this excludes those people who are landless and are unable to benefit from the possibilities of owning a piece of land. Unfortunately, this is a large number of people.

Why are people landless? The original cause of landlessness was at the beginning of the twentieth century when the British settlers took over half of the arable land in Kenya. Anyone who remained on the land became "squatters" with no right to their former land whatsoever. The settlers needed labor so they were willing to allow a certain number of people to remain on the land as squatters. They could be evicted at any time, performed manual labor with low compensation, and consequently were unable to prosper. There are now over ten times the number of people in Kenya than at the beginning of the last century so these squatters and their descendants have increased substantially. There are estimates that over five percent of the Kenyan population are squatters living in rural areas. This is more than two million people.

These are not the only landless. When a women's husband dies, the wife and her children are frequently ousted from the husband's plot by the brothers-in-law. The woman can return to her father's place, but she traditionally has no right to land there. If she has girls, they may perhaps marry a man who has access to land, but her sons will be landless.

Squatter houses on our way to Eldoret. Notice the simple construction of mud and wattle with grass roof and very small plot.

Other people leave their homestead because of family disputes. The media is full of conflicts among family members over land. Those that make the media are frequently due to violence because of land. The losers have to leave and join the landless.

In some cases the land has been subdivided into such small plots that each family cannot do much more than build a house on their allotment. This is particular the case from Gladys' home area, Viyalo, which has been overpopulated since the 1920s. As a result many of Gladys' relatives and neighbors have purchased land elsewhere including many near Lumakanda. Some of those who did not have the resources to purchase land have moved to Nairobi or other main towns, frequently living in the slums.

One positive attribute of small-scale Kenyan farmers is that they will not borrow money using the security of their land as collateral. Consequently, unlike India, for example, people do not lose their land to the moneylenders.

Charles Kunukha, the leader of 1,500 squatters in Mautuma, 15 kilometers (10 miles) from our house is shown speaking last year to the squatters. They had invaded the Lugari forest and in 1994 the Kenyan government promised each family five acres. Yet they have been given only two acres each. They are demanding title deeds for their plots since they are afraid the government will expel them from the forest as it has in other places in Kenya where squatters have invaded forest land.

The landless well realize the benefits of owning land and are desperate to acquire it. They move on to any land that seems vacant or unused. For example, one of the major contentious problems in Kenya is landless people who seize land in the forests at the top of the seven water towers in Kenya. As they cut down the forest, rather than being absorbed in the ground leading to flowing streams used by people downstream, the rain rushes downhill and the streams dry up in the dry season. For example, Transforming Community for Social Change has dealt with farmers on Mt. Elgon who were diverting all the water in a stream so that people lower down from the mountain would have no water during the dry season. This has a potential for violence as the waterless people are desperate. The fact that these two groups are different tribes makes the situation particularly prone to violence.

A picture of Kibera slum in Nairobi with an estimated population of 250,000 people. Notice on the horizon the large office buildings of Nairobi's central business district.

Many of the landless are forced to move to the slums in Nairobi and the other major cities and towns. It is estimated that up to sixty percent of the four million Nairobians live in its more than 200 slums. This is at least two and a half million people. Other cities and towns also have substantial slums. While many consider this move from the rural countryside to urban areas as a positive movement, in the case of Kenya, I see this as a very negative development as more and more people are squeezed into the current slums and newly developing ones. These people mostly live through day labor when they can get it. Life is precarious. They spend fifty percent or more of their income on food and, when the price of food rises as it did in 2017 during the drought, their lives become even more precarious.

I estimate that over twenty-five percent of the Kenyan population is landless. This is one of the major problems in Kenya leading to poverty, unrest, and high levels of violence. While there is much talk about this problem, the solutions are difficult to implement, so little is accomplished by either the government or non-government organizations. These solutions are merely band-aids rather than addressing the roots of the problem. The government sometimes buys a large farm and divides it up into small plots for those who have been evicted elsewhere. This is usually quite contentious because the local people near the large farm feel that "foreigners" (i.e., people from a different tribe or clan) are invading their area.

As I have indicated, small-scale farming produces higher yields than large commercial farms. I recommend that a new Million Acre Settlement Scheme be implemented. Due to increased need, the average plot should be about five acres rather than the twenty acres in the original scheme. If done properly, this could accommodate one to two million landless people.

American Exceptionalism

A present day hog establishment in the United States.

American exceptionalism usually refers to that amazing form of government established at the end of eighteenth century by a group of white males, many being slave owners, that made the United States, well, "exceptional" and superior to all other countries in the world. More formally Wikipeida defines it as "an ideology holding the United States as unique among nations in positive or negative connotations, with respect to its ideas of democracy and personal freedom. ... Another theme is the sense the United States' history and mission give it a superiority over other nations."

Why would I be writing a Report from Kenya about small-scale farming on American exceptionalism? The reason is that this is one of the major factors destroying small-scale farming in Africa and elsewhere.

I have found that many Americans consider, not just the form of government, but everything Americans do as superior to that done by other people in the world. I have observed this attitude time after time in American visitors to East Africa. When talking with Africans about issues, including agriculture, their conversation is based on the assumption that what America does is so much better than what Africans so, and so everyone in the world should adopt American methods. This includes agricultural development, that is, the replacement of the small-scale farmer with large commercial farmers using mechanization, hybrid seeds, fertilizers, insecticide and herbicides bought from gigantic companies. This is always couched with the rationalization that this is needed to feed the 9 to 10 billion people who will be living in the world in 2050.

A good number of years ago, the Rocky Mountain Institute released a study on the amount of methane cows produce. They took as their baseline the amount of methane that a grain-eating American cow produced. Cows are not normally grain eaters, so the amount of methane they produce is increased by the abnormal grain fed to them in feedlots. They then estimated the number of cows in the world and concluded that cows are responsible for a high percentage of the methane produced in the world. I sent them a letter to the editor stating that cows in Africa are only one-third to one-fourth the size of American cows and moreover are grass-fed so don't release methane from being fed grain. I then suggested that their total was perhaps four to five times too high. They did not publish my letter.

On a simpler level take bacon. When grain-fed bacon is fried in the United States, the resulting grease has to be poured out of the frying pan. Here in Kenya, bacon is so lean, that oil needs to be put in the frying pan in order for the bacon not to stick. Why should the world adopt the unhealthy American bacon? Wouldn't it be better if Kenyan farmers taught the American how to raise lean, healthy bacon and pork?

This is a picture of a house being built on the side road to our house. About three or four months ago the owner put a fence around the property which has kept the cows, goats, and sheep from eating the grass. The grass inside has grown quite tall in that short time. Notice, though, that the grass on the road is short because it has been grazed by livestock. Therefore there is no need to mow the grass with polluting, noisy lawn mowers as is done in the United States.

Muslims don't eat pork. Hindus don't eat beef. Unlike some other people, American don't eat horses or dogs. Kenyans don't eat donkeys, but people in the Middle East do eat donkeys. This has led to two problems in Kenya as donkey abattoirs have opened to sell donkey meat to the Middle East. First donkeys are now being stolen to be sold to the abattoirs. Second with over 1000 donkeys being slaughtered per day, there is fear that the abattoirs are decimating the donkey population in Kenya. The prize meat in Kenya is called in Swahili, *myama choma*, roasted goat meat. Since goat meat is the healthiest red meat and goats are browsers and are not fed on grain or growth hormones, I do not understand why Americans, the heaviest meat eaters in the world, are so adverse to eating goat The average consumption of goat meat in the US is one quarter pound per person per year.

American commercial farming has many problems including soil erosion, pollution of air and water, unsustainable use of ground water, benefits going to large corporations rather than the farmers themselves, corn for ethanol and animal feed, and so on. There are many articles on the Internet that discuss the many problems of American agriculture. I will not discuss these major problems since a search will give the reader more than enough articles to read. American "experts", in their worldwide evangelism of the wonders of American agriculture, promote large commercial farming in place of indigenous agriculture. Given the problems with American agriculture, why should it be promoted in other countries of the world?

Donnie Smith, the former chief executive of Tyson Foods, the second largest food processor in the United States, is spearheading a gigantic scheme in Tanzania on a 500-acre plot on the foothills of Mt. Kilimanjaro. The

The broiler chickens that Tyson Foods plans on bringing to East Africa.

goal is in a few years to export 40 million to 50 million chicks per year to the sounding countries. According to Smith "It's God's plan". He feels "impelled by his faith to feed the world's poor." The plan is to use expensive maize (corn) that is used to feed people to now feed chickens, so that the middle class can have a tasteless chicken for dinner. This will undoubtedly raise the price of maize so that the poor in those countries will pay for those chicken dinners. I predict that this scheme will mostly fail <theguardian.com/environment/2018/dec/27/its-gods-plan-the-man-who-dreams-of-bringing-intensive-chicken-farming-to-africa>.

Americans think that they "know it all" and whatever they promote is the ultimate answer for world agriculture everywhere. If this were just academic, there would be no problem, but the American experts have two major advantages. The first is that American government organizations such as US Agency for International Development (USAID) and many non-government organizations have lots of money, resources, well-educated staff who see it as their calling to introduce American agricultural methods to the African farmers. State department officials posted at US embassies are part of this lobby group that controls the conversation.

The second advantage is that American corporations like Monsanto and Tyson Foods have the funds and interest to promote their products. This includes placing articles in the Kenyan papers by Western-educated Africans, frequently in their pay, about the virtues of their products. Over the years I have watched as American companies have promoted GMO crops in Kenya as the answer to enhanced crop production. After many years of GMO crops promotion, Kenya has finally agreed to allow GMO cotton crops to be planted. This is supposed to be a testing phase, but since the GMO companies will assess the results, I think the conclusion has already been determined.

The saving grace—it is interesting how I am using religious terminology, but this is because American exceptionalism is a secular religion—is that the vast majority of attempts to bring American commercial agriculture to Africa fail, sometimes spectacularly. Yet if the resources and efforts put into these failures were channeled into improving small-scale farming in Africa, the results would increase crop production that would benefit a large number of Africans.

The Demographic Transition in Kenya

When I lived in Kenya in the 1960s, Kenya had a fertility rate (the number of children the average woman would have during her child-bearing years) of over eight children per mother. At that time this was considered the highest birth rate in the world. Infant and child mortality rates were also high. This high rate was obvious as most women of child-bearing age were either pregnant, with a baby on their back, or both. I knew one of the two women in Machakos District who was providing family planning for a population of a million people. With scarce resources, they were overwhelmed by the need. By the time I left Kenya in late 1970, Kenyans began to become alarmed by this abnormally high birth rate and the birth rate slowly began to decline.

This decline led to a fertility rate in the 1990s of 4.5 children per woman of childbearing age. Therefore the "mythical" average family at that time had the two grandparents, four sons, four daughter-in-laws, and eighteen grandchildren. This is a large family with a total of twenty-eight people. This meant that there was a large labor pool to cultivate the small-scale farm. With this large pool of labor it was beneficial for one or more of the family to seek outside employment to help support the family and raise capital for input into the farm.

In 2018 the Kenya fertility rate (Source: CIA Facybook) is 2.81 children. With a lower but still fairly high infant, child, and maternity death rate, the replacement level is 2.3 children per mother. The rate is dropping quickly so that the replacement level should be reached in 2020 or 2021.

It has taken about 50 years—1970 to 2020 —for this demographic transition to occur in Kenya. This is extremely fast and probably the fastest in Africa to date. I don't understand why the demographers have not been noticing this trend and analyzing why it has occurred. I predict that like many countries, once the replacement level in Kenya has been reached, it will continue to go down below replacement level.

Some of our family - in back, grand-niece Trinah; center left, grand-son Brian; center right, grand-daughter Faith (with shirt given to her by my grand-daughter Jayla), and front, grand-daughter, Rembo.

This means that the current family has two grandparents, 2.25 sons, and 2.25 daughter-in-laws, and about 5 grandchildren for a total family unit of 11.5 people. Of course the size of the farm of these grandparents is one-fourth of that of their parents. The family is still large enough so one or more members may go off to paid employment while the rest will tend to the family farm.

If we project this into the next generation in 2050, since the replacement level will have been reached a long time previously, the average family will then have two grandparents, one son, one daughter-in-law, and two children for a total of only 6 people. While labor will now be in short supply, land will no longer be divided into smaller plots so the son will have the same size plot as his father. The family plot will now be about 40 percent of the parent's plot and only 10 percent of the grandparent's plot at the time of independence in 1963. This will bring tremendous change to the dynamics of small-scale farming. While I hope to be around in 2050 to see how this turns out, I'll be 107 years old by then so I may not be able to write you an update.

While this all may seem academic, it parallels the trends in Gladys' family. Her mother had ten children of whom seven survived to adulthood. They happened to be all women and they had 7, 6, 5, 3, 2, 2, and 0 children for a total of 25 children, an average of 3.6 each. Twelve of these are females. Most of them are still in their childbearing years, but the oldest are near the end. So far none of them has more than two children, although some may give birth to more children than they have now. Nonetheless I doubt that the total will be more than the 27 children needed for replacement. Gladys and all five of her living sisters take care of one or more grandchildren and one sister is now taking care of a great-grandchild, while the fathers and mothers are seeking their fortunes in Kakamega, Eldoret, Nairobi, or elsewhere. Some of them have found it while others are still seeking. Some of our more successful nephews have already built or are beginning to build a house in their parents' rural homesteads or on land they have purchased nearby.

Reasons for the Demographic Transition in Kenya

Kenya Population - 2019

Age	Total	Female	Male
Total	49,142,516	24,630,484	24,512,032
0-4	5,531,359	2,752,479	2,778,880
5-9	6,473,488	3,228,365	3,245,123
10-14	6,650,319	3,318,340	3,331,979
15-19	5,605,722	2,802,364	2,803,358
20-24	4,292,846	2,153,952	2,138,894
25-29	3,988,032	1,995,127	1,992,905
30-34	3,796,346	1,891,436	1,904,910
35-39	3,402,750	1,690,216	1,712,534
40-44	2,600,659	1,262,969	1,337,690

Source: <geoba.se/country.php?cc=KE>

This is a chart of age groups in five year increments. The number of women in their fertile years increases substantially from the older women through the younger women. The number of children born in the last fifteen years declines with each 5-year age group. This indicates that even though there are more women of child-bearing age, they are having significantly fewer children. Replacement level is near.

Nonetheless I don't see that Kenyans are going to follow the American example from the twentieth century where people in rural areas abandoned their homes for life in the cities. Kenyans may continue to go to work in the cities, but I expect they will continue to invest in their homes in the countryside. After their working days are over, who wants to stay in the slums in the cities or the small apartments in high rises? Many retired Kenyans that I know have already moved back to the countryside so I expect this trend will continue. Consequently small-scale farming will also continue, perhaps more intensively because the farms are smaller and perhaps with more mechanical efficiency as labor becomes scarcer and more costly.

What is happening in Kenya is positive, but the demographic transition is coming very slowly to Uganda, Burundi, Tanzania, Democratic Republic of the Congo, Ethiopia, South Sudan, and Somalia. Rwanda may be an exception as a quicker demographic transition may be occurring there, but what is happening in Kenya is unique and so far not a model for other nearby countries.

My posting last week on the demographic transition in Kenya has resulted in a much greater response from readers than most of my posting. Some people have asked why this has occurred in Kenya. Naturally the reasons are complex, but I can give you some ideas.

1. *More Education for Girls*: While it is common to say, when girls receive more education, their fertility rate declines, but I am not so sure. In Uganda where women are becoming more educated, the fertility rate is going down very slowly. Also, many primary and secondary school girls get pregnant. Tanzania has just announced a rule that girl students who have become pregnant must be expelled from school and not allowed to return. I think that the correlation between schooling and fertility rate

is different. As all kids go to school, now in Kenya everyone is supposed to finish secondary school. Except for weekends and holidays, children are no longer a major source of labor for the family farm. Although schooling is "free", the parents incur considerable expenses for uniform, books, supplies, fees that schools charge, and so on. I think parents realize that they can't afford to provide for lots of kids who are going to school.

2. *The HIV+ Rate for Women in Kenya is 5.2 percent.* HIV+ women are advised not to become pregnant. While some may still give birth, I doubt that they will have large families.

3. *Infertility*: One of my readers commented that perhaps the modern lifestyle leads to higher infertility rates among women. The significant increase in the rates of diabetes, high blood pressure, and cancer in the current population is supporting evidence for this possibility.

4. ***End of Female Genital Mutilation (FGM):*** Before the Europeans arrived, 75 percent of the Kenyan women were "cut" as they say here. The missionaries were horrified and opposed the practice for anyone who became a Christian. The Luo and Luhya (the tribe of most Kenyan Quakers) did not practice FGM. By about 2000 the number of circumcised women was down to something like 46 percent. Beginning in 2006, Kenya began a strong effort to stamp out the practice including making it illegal. Since girls were circumcised at about 14 and then were eligible for marriage, usually to a much older man, this would give rise to very large families. The rate is now supposed to be under 20 percent of all adult women and declining rapidly for the younger generation. Uncircumcised girls will marry much later which will reduce the number of children they will have.

5. ***Abortion:*** Except to save the life of the mother, abortion is illegal in Kenya. Nonetheless there are over 464,000 (2012 figure) abortions per year or an abortion ratio of 30 per 100 live births. Yearly 120,000 women receive care in health facilities for complications of induced abortion; Kenyatta Hospital in Nairobi has a special ward just to serve women with botched abortions. At least 2,600 women per year die from abortions. Many others will become infertile.

6. ***The Maternity Mortality Rate is still 510 per 100,000 live births***, which totals 6,000 women per year. The US rate is 14 per 100,000 live births. I have gone to funerals of family members who died in childbirth. When a mother dies, or for any other reason the birth mother is not able to

take care of her children and her children are adopted by other family members, the new "mother" will count these children as her own, which could skew the statistics by incorrectly increasing the fertility rate.

7. ***The Infant Mortality Rate*** now 36.1 per 1,000 births, still high but nothing like in the 1960s when there were over 100 deaths per 1,000 births. Research has shown that when the infant mortality rate declines so does the birth rate. There is a strong emphasis in Kenya now for women to give birth in hospitals and clinics. This will decrease the infant and maternal mortality rates.

8. ***New Methods of Birth Control*** are important. At the Ntaseka Clinic in Burundi, almost 81 percent of the women chose contraceptive injection for their preferred method of birth control. Another two percent chose implants. These methods are popular because the women don't have to tell their husbands/boyfriends that they are on birth control. The women themselves are therefore making the decision. The morning after pills is also important. I know of a young woman who was raped during a robbery and the hospital gave her the morning after pill.

9. ***Prevalence of Birth Control:*** While on the whole, birth control methods are available throughout the country, there are some parts of the country where they are scarce, difficult for young unmarried women to access, or out-of-stock at medical facilities. The latest information is that 61.6 percent of women of child-bearing age are using some method of birth control.

10. ***Media:*** TV and other media in their fashion sections show glamorous young women with no children depicted. They clearly show that having children is not the prevalent culture. Even the soap operas on TV watched by many women have very few children depicted in the family dramas.

11. *Cell Phones* came into general use beginning in 2007. Now most Kenyans have a cell phone or access to a cell phone. This has opened up people's horizons as they are no longer isolated in their home community. They are more cosmopolitan. This works against the former concept that the main purpose of a woman is to give birth.

12. *Smart Phones* have become quite common in the last few years. While I am certain the moralists will be horrified by this thought, but anyone with access to a smart phone has access to pornography. I have now begun hearing preachers complain that the youth watch too much pornography. But pornography upends the concept that the purpose of sex is to procreate as it promotes the concept that sex itself should be pleasurable. This change in the attitude towards sexuality lowers the expectation that women should have lots of children.

13. *Affluence:* Unlike the other countries in eastern Africa, Kenya is now considered a lower middle income country by the World Bank. I don't think that money itself is a significant factor, but money provides the resources to buy TVs, smart phones, and print media which support many of the issues mentioned above.

David Zarembka (1943-2020)

David Zarembka, originally from Clayton, Missouri, died on April 1, 2020 in Eldoret, Kenya, after being hospitalized for COVID-19. David was born May 6, 1943, to the late Richard Zarembka and the late Helen Jane Colvin Zarembka. David graduated from Clayton High School in 1961 before enrolling at Harvard University. In 1964, after his junior year in college, he took a year off to teach Rwandan refugees in western Tanzania. He then returned to Harvard and graduated with a Bachelor of Arts in African History in 1966.

After graduation, he joined the Peace Corps, serving in Tanzania and Kenya for two years. David then became the founding Head of the Mua Hills Harambee Secondary School in Machakos District, Kenya. He married Rodah Wayua Malinda in 1969 and returned to the United States in 1971, where he received his Masters of Arts in International and Development Education from the University of Pittsburgh. David and Rodah had two children, Joy and Tommy Zarembka, while living in Pittsburgh, Pennsylvania. They joined the Religious Society of Friends (Quakers) and became active members of the Pittsburgh Friends Meeting. David founded a number of organizations, including an alternative high school, a retail food co-op, a wholesale food co-op, and a housing co-op.

In the late 1980s, David moved to Yellow Springs, OH, and later the Washington, DC area. He pursued his humanitarian mission for peace and social concerns, doing home repairs to finance his dreams. He joined the Bethesda Friends Meeting and, in 1996, unofficially adopted Douglas Kebengwa. In 1998, David founded the African Great Lakes Initiative (AGLI) of the Friends Peace Teams, an organization dedicated to peacemaking activities in Eastern and Central Africa. Through AGLI, he introduced the Alternative to Violence Project and helped develop the Healing and Rebuilding Our Communities (HROC) program in Rwanda, Burundi, the Congo, and Kenya.

David was remarried in 1999 to Gladys Kamonya and moved back to St. Louis to be closer to his ailing mother. In 2007, they moved to Lumakanda in western Kenya to continue his social justice and peace work, which inspired his book, *A Peace of Africa*, published in 2011. They became members of the Lumakanda Friends Church. David was instrumental in initiating the Friends Women's Association Kamenge Clinic in Bujumbura, Burundi, an orphans program and technical school in

Bududa, Uganda, and a summer work camp program in the region. He also spearheaded Quaker speaking tours throughout Europe, Africa, and the Americas for African peace experts. He was the editor of the AGLI publication, *PeaceWays AGLI* from 1998-2015 and subsequently continued the work through his "Reports from Kenya" series (Google Group "Reports from Kenya") and numerous journal publications.

Those left to cherish and honor his loving memory include: first wife, Rodah Wayua (Mua Hills, Kenya); daughter Joy Zarembka (Washington, DC), son Tommy Zarembka (Gaithersburg, MD), and son Douglas Kebengwa (Hyattsville, MD); his grandchildren Naomi, Matias and Jayla Zarembka in the USA; his extended family of grandchildren in Kenya; his siblings Paul Zarembka (Buffalo, NY), Elaine Belmaker (Modiin, Israel), Arlene Zarembka (University City, MO), Michiko Shimizu (Tokyo, Japan); and in-laws, nieces, nephews, and cousins in the USA, Kenya, Israel, and Japan. In loving memory, we honor his second wife of 21 years, Gladys Kamonya (Lumakanda, Kenya) who succumbed to COVID-19 a week prior to David's death.

In lieu of flowers, please donate to these organizations that David held so dear: Transforming Community for Social Change (Kenya); Friends Women's Association (Burundi); Innovations in Peacemaking (Burundi); and Healing and Rebuilding Our Communities (Rwanda).

Services: David's heart was always in Africa. With a burial service on April 7th, 2021, his body was laid to rest in Kenya next to his beloved Gladys Kamonya. A virtual memorial service celebrating Gladys and David was held by the St. Louis Friends Meeting on May 8, 2021, to celebrate their life and love. An additional in-person memorial service by the Lumakanda Friends Church of Lugari Yearly Meeting is planned when it is deemed safe.